Stewardship

Biblical Foundation

By Steven P. Demme

1-888-854-MATH (6284)
MathUSee.com

Math·U·See

1-888-854-MATH (6284) - MathUSee.com

Printed in the United States of America.

Stewardship | BIBLICAL FOUNDATION

BIBLICAL COPYRIGHTS

INTRODUCTION

Believing Bereans Who See Jesus

This book began as a means of teaching my four sons (all now in their twenties) what God had taught me about having my own business and handling money. I hope it is also helpful to you. It is not meant to be the definitive word on finance in the scriptures. Rather, it is my testimony, and my wife's, of how God has led and taught us from His Word in the everyday experiences of life for the past quarter of a century. I am going to give many specific scriptures and how they have been applied in my life. You may not agree with my conclusions, but I hope we can all be Bereans and search the scriptures to see whether what I have written is true.

"Now these were more noble than those in Thessalonica, in that they received the word with all readiness of the mind, examining the Scriptures daily, whether these things were so" *Acts 17:11, ASV.*

It is my earnest hope that this collection of devotionals will stimulate discussion between parents and children as they consider issues of finance and examine them in the light of the eternal principles found in God's Word.

If you think of a topic you wish I had covered, or a specific insight that God has given you, please communicate with me via email. I can be reached at steve@mathusee.com.

If you would like to be kept abreast of new developments

and additions to this book, please sign up at www.mathu-see.com/stewardship.

Let me share an additional note that my sons have taught me. For those of you who are teaching your children biblical principles, be careful that you make a distinction between your opinion and God's opinion as taught in the revealed Word of God. May I remind you how easy it is to misuse our authority as parents as a bully pulpit for our own ideas, when it is God's ideas we should be pursuing. May I encourage us each to read, pray, and seek God for those principles that will endure. Let's take our children to God, and His Word, for guidance that will never disappoint.

"Heaven and earth shall pass away, but my words shall not pass away" *Matthew 24:35.*

As you study God's Word, I hope you will not only learn more about biblical principles, but also about God Himself. There are over 1,000 references to money in the Bible. Many of them are located in the book of Proverbs. The aim of *Stewardship* is to help you learn and apply wisdom to the study of our personal finances. It is not about how to make more money, how to develop a personal fortune, or how to make your first million before you are 30. The focus is to learn to see money and think about finances the way God does and to be transformed by the renewing of our mind, so we are not conformed to the world's way of looking at this critical area in our lives.

As you read Proverbs and learn about wisdom, let me encourage you not only to see the biblical principle, but also to see Jesus. Proverbs 8:22–32 helped me see that Jesus is wisdom in the flesh. When I read this passage, I hear

Jesus talking of Himself and His relationship with His Father, closing in an exhortation to us as His children to follow Him and keep His ways.

"The Lord possessed me in the beginning of his way, before his works of old. I was set up from everlasting, from the beginning, or ever the earth was. When there were no depths, I was brought forth; when there were no fountains abounding with water. Before the mountains were settled, before the hills was I brought forth: while as yet he had not made the earth, nor the fields, nor the highest part of the dust of the world. When he prepared the heavens, I was there: when he set a compass upon the face of the depth: when he established the clouds above: when he strengthened the fountains of the deep: when he gave to the sea his decree, that the waters should not pass his commandment: when he appointed the foundations of the earth: then I was by him, as one brought up with him: and I was daily his delight, rejoicing always before him; rejoicing in the habitable part of his earth; and my delights were with the sons of men. Now therefore hearken unto me, O ye children: for blessed are they that keep my ways" *Proverbs 8:22–32.*

As you seek to understand wisdom, seek also for a fresh glimpse of Jesus within the book of Proverbs. I chose Proverbs 3:13–18 to introduce us to Proverbs because I believe it sets the tone for the rest of the book. My comments follow the verses.

"Happy is the man that findeth wisdom, and the man that getteth understanding" *Proverbs 3:13.*

Another word for happy is blessed. You will be blessed and happy when you acquire wisdom, and Jesus is wisdom in the flesh. He is the Word incarnate.

"For the merchandise of it is better than the merchandise of silver, and the gain thereof than fine gold. She is more precious than rubies: and all the things thou canst desire are not to be compared unto her" *Proverbs 3:14–15.*

Wisdom is better than things, and that includes money.

"Length of days is in her right hand; and in her left hand riches and honour. Her ways are ways of pleasantness, and all her paths are peace" *Proverbs 3:16–17.*

The benefits of following after wisdom and/or following Jesus are a long life, riches, honor, and peace.

"She is a tree of life to them that lay hold upon her: and happy is every one that retaineth her" *Proverbs 3:18.*

What great promises for the one who believes in and receives Jesus! Those who lay hold of Jesus and retain Him will soon find His blessing following along as day follows night. A tree of life and a blessed, happy life.

God encourages us to ask. At the end of each lesson, you will find a scripture or a prayer. I hope these will be a useful aid for you in asking for, and connecting with, the grace of God.

"And I say unto you, Ask, and it shall be given you" *Luke 11:9.*

". . . Ye have not, because ye ask not. Ye ask, and receive not, because ye ask amiss, that ye may spend it in your pleasures" *James 4:2–3, ASV.*

"And this is the boldness which we have toward him, that, if we ask anything according to his will, he heareth us: and if we know that he heareth us whatsoever we ask, we know that we have the petitions which we have asked of him" *1 John 5:14–15, ASV.*

These scriptures have marked my life lately.

Recognizing that God is the source of all I need, I have found myself asking more and more, and receiving more and better. As the hymn says, "All I have needed Thy hand hath provided."

As you read these short devotions, if there is any area where you sense a need for God's help, simply ask God to meet your need, but ask according to His will. We are assured that our good God, who desires only what is best for us, will do what we ask, as long as it is according to His will. And we know His will because we have His revealed Word, the Bible.

For example, in lesson 1 we are considering that we do not own anything. The earth is His, and I need to be reminded of this. We can apply this principle and pray something like this: "God, by Your good Spirit, remind me daily that all the things I have are from You. Help me to be a wise and faithful steward of the resources at my disposal."

This is a part of what it means to be a lifelong learner or disciple. When we hear something that we want to put into practice, we commit it to the Lord and trust Him to make it a part of our lives. We ask Him to write it on our hearts so we will be doers of the Word and not hearers only.

"Ye are an epistle of Christ . . . written not with ink, but with the Spirit of the living God; not in tables of stone, but in tables that are hearts of flesh"
2 Corinthians 3:3, ASV.

"But be ye doers of the word, and not hearers only"
James 1:22.

PRAYER

Give us the same spirit as those saints in Berea who received the Word with all readiness of the mind, and examined the scriptures daily, to see whether these things were so. Cause us to be hearers and doers of Your Word. Enable us to see finances from Your perspective, and give us grace to apply what we learn in our day-to-day dealings with money. And please open our eyes to see Jesus in all we learn and do. Amen.

Note: Unless otherwise noted, the scripture references are from the King James Version of the Bible.

LESSON 1

Faithful Steward

"Know therefore this day, and consider it in thine heart, that the Lord He is God in heaven above and upon the earth beneath: there is none else" *Deuteronomy 4:39.*

"And God blessed them: and God said unto them, Be fruitful, and multiply, and replenish the earth, and subdue it; and have dominion over the fish of the sea, and over the birds of the heavens, and over every living thing that moveth upon the earth" *Genesis 1:28, ASV.*

STEWARD

In the beginning, God. That is where any study, of any subject, begins and ends. He is the Creator and we are His creation. He is the potter and we are the clay. In Romans, the first chapter, we read of those who refused to have God in their knowledge (v. 28). May this never be said of us. On the contrary, we do want God in our knowledge and in every area of our lives. The best way I have found to have God in your knowledge is to read God's Word regularly. In scripture, we find that we really don't own anything, God does. Instead of protecting our own possessions, we are really taking caring of His "stuff." We are not owners but stewards. Here is a definition of a steward as found in Noah

Webster's 1828 dictionary:

> **Steward** *n.* A man employed in great families to manage the domestic concerns, superintend the other servants, collect the rents or income, keep the accounts, etc.

As Christians, we are employed by the greatest family, the family of God, to manage the resources He has placed in our care. We are working side by side with a multitude of fellow believers to manage the affairs for the great God, the King of kings, who owns the cattle on a thousand hills *(Psalm 50:10).* At creation, God called His people to "replenish the earth, and subdue it; and have dominion" *Genesis 1:28.* It is our calling to be stewards and not owners.

My wife's and my experience immediately after marriage helped us with this concept of stewardship, as opposed to the idea of ownership. Our first domicile was in the back of a large New England house, up a long flight of stairs in the former servants' quarters. After four months there, we moved into a church parsonage in Georgia, where we lived for almost nine years. Then, after living in two more rentals and being married for over 12 years, we were able, with my parents' help, to purchase our first home. By that time, most of the wanting to "have our own place" had drained out of us. When we were younger, we dreamed of this, but after 12 years of living in someone else's home, most of our desire for a place of our own had dissipated.

When we finally bought our first home, we had an evening service with friends from the local church. There we renewed our desire to commit all of our belongings to God, formally dedicating the house to God and giving it over to Him for His purposes. All the furnishings, vehicles, and possessions were rededicated to Him.

"The earth is the Lord's, and the fulness thereof; the world, and they that dwell therein" *Psalm 24:1.*

I don't know how real this concept is to each of you, but I trust it will grow in your understanding. It makes a huge difference when you recognize that you really don't own anything.

In many ways, it is easier knowing that all of your possessions are His. When you live in a rental property and you have a leak in your roof, you just call the landlord and he takes care of it. Someday you may have your own home. Maybe you'll find a leak in the roof. When you remember that your house is really His, call on your heavenly landlord and let Him know He has a problem with His roof. Ask Him how He intends to fix it! It really is a liberating way to live. I confess I have difficulty remembering this concept in every crisis, but when I do, it is comforting to know that the problem is His responsibility.

There is a great story written about this very concept called *The Pineapple Story*. It is published by IBLP. You can find it at http://store.iblp.org.

FAITHFUL

"Here, moreover, it is required in stewards, that a man be found faithful" *1 Corinthians 4:2, ASV.*

God values faithfulness. The dictionary defines it as "steady in allegiance or affection; loyal; constant: *faithful friends.*" It is such a vital concept to grasp that I am going to list several synonyms to help you understand this concept more fully: loyal, steady, solid, enduring, dutiful, constant, reliable, firm, steadfast, true, resolute, staunch, devoted, firm, trustworthy, unwavering, dependable. The phrase in scripture that comes to mind when I meditate on faithfulness is "patient continuance in well doing," found in Romans 2:7.

The writer of Proverbs implies that a faithful man is rare and hard to find. "A faithful man who can find?" *Proverbs 20:6.*

Faithfulness is the very nature of God. "God is faithful" *1 Corinthians 1:9.*

Jesus is called faithful and true. "Behold, a white horse, and he that sat thereon called Faithful and True; and in righteousness he doth judge and make war" *Revelation 19:11, ASV*. Abraham and Moses were described as faithful men. "My servant Moses is not so; he is faithful in all my house" *Numbers 12:7, ASV.* "The faithful Abraham" *Galatians 3:9, ASV.*

When you couple these two words, *faithful* and *steward*, you have the essence of what a Christian is to be, especially in regard to money and possessions. We recognize that all that we have belongs to God and we are simply stewards of what He has placed in our care. Then we trust God to make us faithful in using these resources to provide for our families and extend God's kingdom on this earth.

PRAYER

God, by Your good Spirit, remind us daily that all the things we have are from You. Help us to be wise and faithful stewards of the resources at our disposal. Amen.

LESSON 2

The Root of All Kinds of Evil

When we think of money, one of the first scriptures that normally comes to mind is "For the love of money is a root of all kinds of evil: which some reaching after have been led astray from the faith, and have pierced themselves through with many sorrows" *1 Timothy 6:10, ASV.*

Notice, money is not evil, the *love* of money is. So why is the love of money such a stumbling block? You have to look at the study of money as you would any other subject, examining it as children of God and as a part of God's family. As Christians, our first and highest calling is to love God with everything that is in us.

"And thou shalt love the Lord thy God with all thy heart, and with all thy soul, and with all thy mind, and with all thy strength" *Mark 12:30.*

Anything that takes away from or diminishes our devotion to Christ has to be reexamined. Jesus is our wonderful Master. He is our Lord. We cannot love or serve two masters, nor do we want to. He is loving, good, and caring.

"No servant can serve two masters: for either he will hate the one, and love the other; or else he will hold to the one, and despise the other. Ye cannot serve God and mammon" *Luke 16:13.*

Money has the potential, as few things do, to replace God in our heart of hearts. It can become an idol. An idol is not restricted to a wooden figurine or a bronze statue. An idol is anything that we look to in order to supply what only God can provide. Money can provide a sense of security. It can provide food, clothing, and shelter. Money can be a source of pride, and acquiring more and more of it may become a lifelong passion. Working for money, and seeing it as the provision for all of your needs, is making money an idol. Left alone, this desire for money has the potential to become your master.

It is always a temptation to serve what we can see and touch. Throughout the history of God's people in the Old Testament, there was always this conflict between serving and worshiping God and worshiping idols. The names of the idols may have changed, but the same tendencies are ever-present in the human heart. The Pharisees also had this problem. "And the Pharisees, who were lovers of money, heard all these things; and they scoffed at him" *Luke 16:14, ASV.*

I hope you will be free from the love of money. Let me encourage you to love God with all your heart, soul, mind, and strength. If money comes your way, be grateful for it. It is not to be feared; it is to be used for building the kingdom of God on this earth. Hold it lightly then, so if Jesus calls you to sell all you possess and give it to the poor, you can do so gladly. Pray that you won't have the same response as the young man in Matthew 19.

"Jesus said unto him, If thou wouldest be perfect, go, sell that which thou hast, and give to the poor, and thou shalt have treasure in heaven: and come, follow me. But when the young man heard the saying, he went away sorrowful; for he was one that had great possessions" *Matthew 19:21–22, ASV.*

Of course, as my brother is fond of saying, it is easier to lay all of your possessions on the altar when you are young. After all, what do you have but a used car, some clothes, maybe a computer, and a few books? It does get tougher the more things you acquire. But I do hope that no matter how many possessions you acquire, or how many blessings you receive, you will always remember that it is all from Him and for Him, for He is God.

Money is not a bad thing, in and of itself. It has the potential for great good. It is like fire: it can burn, or it can provide heat. There is nothing intrinsically wrong with money, any more than there is with fire. It depends on your attitude toward it and how you use it. Money and possessions can be used for extending God's kingdom to the uttermost parts of the earth, as well as for feeding the poor in your neighborhood. Money is a tool to be used for His glory. Let us see it as a gift from Him and be good stewards of it.

PRAYER

Lord of heaven and earth, help us to love You with all that is in us. Save us from the love of money. Root the desire for money out of our hearts. Enable us to see the resources that You place at our disposal as gifts from You, and give us the wisdom to use it for Your glory. Amen.

LESSON 3

Loving God or Coveting

Loving is giving without any expectation of return. It is thinking of others first. Remember the acronym JOY: Jesus first, Others second, Yourself third. God Himself loves us by giving.

"For God so loved the world, that he gave his only begotten Son, that whosoever believeth in him should not perish, but have everlasting life" *John 3:16.*

It is only with God living in us that we can expect to learn the art of loving others and putting them first instead of ourselves. We are, by nature, selfish. Even the few philanthropic deeds we do may be done with selfish motives. Pure love and selfless giving come from God alone.

On a personal note, my wife and I have the privilege of having several friends who really love us. Their love is evidenced in the genuine way they rejoice with us when we receive material blessings. The Bible speaks of rejoicing with those who rejoice and weeping with those who weep. Our friends fulfill that scripture. Instead of wishing they had received the blessing (coveting), they are really happy with our good fortune (loving).

When you covet, you are thinking only of yourself. When you truly love another, you are thinking only of him. Coveting and loving your neighbor as yourself are polar opposites.

"Neither shalt thou covet thy neighbor's wife; neither shalt thou desire thy neighbor's house, his field, or his man-servant, or his maid-servant, his ox, or his ass, or anything that is thy neighbor's" *Deuteronomy 5:21, ASV.*

"Thou shalt love the Lord thy God with all thy heart, and with all thy soul, and with all thy mind. This is the great and first commandment. And a second like unto it is this, Thou shalt love thy neighbor as thyself. On these two commandments the whole law hangeth, and the prophets" *Matthew 22:37–40, ASV.*

Loving and coveting are unseen attitudes and are found in the inner man, in the hidden man of the heart. They are hard to discern outwardly. Love is selfless, while coveting is selfish.

"Love suffereth long, and is kind; love envieth not; love vaunteth not itself, is not puffed up, doth not behave itself unseemly, seeketh not its own, is not provoked, taketh not account of evil" *1 Corinthians 13:4–5, ASV.*

This attitude of putting others first is a gift from God. It is the antithesis of the selfish attitude of coveting. May it spread to all of us.

I can't cover all of the bases on this subject, but if you recognize that the seeds of coveting are prevalent in the human heart, you will be forewarned and therefore forearmed. The heart is deceitful above all else and desperately wicked.

Jeremiah 17:9. It is only God who has the power to change us from the inside out.

Jesus said that the whole law and the prophets are built on loving God and loving your neighbor as yourself. When I read the Old Testament, I look to see how this truth is applied to specific injunctions and commands. One example of loving your neighbor as yourself is made clear in the command "Thou shalt not steal" *Exodus 20:15*. Stealing is the outworking of coveting.

One form of stealing is taking land that belongs to your neighbor. Even a cursory reading of Numbers and Deuteronomy reveals how God is careful about boundaries.

Another aspect of stealing applies to personal possessions. "Thou shalt not steal" reverberates through several of the Ten Commandments. We should not steal our neighbor's wife (adultery) or his life (murder). Focusing on self leads to stealing. When we love our neighbor, we won't take what belongs to him.

This principle even extends to a popular figure in literature, Robin Hood. I urge you to beware of the "Robbing" Hood mentality. Taking from the rich is theft, plain and simple, regardless of whether you give it to the poor or not. It may have fancy names like socialism and redistribution of wealth, but it is still theft. It seems to have infiltrated many branches of our government today. Redistribution of someone else's personal property is stealing, regardless of how noble it sounds.

Loving thinks of the other and coveting thinks of self. Jesus was and is the most loving being to grace our planet. May we be more like Him.

"Even as the Son of man came not to be served but to serve, and to give his life as a ransom for many" *Matthew 20:28, ESV.*

"Not looking each of you to his own things, but each of you also to the things of others. Have this mind in you, which was also in Christ Jesus: who, existing in the form of God, counted not the being on an equality with God a thing to be grasped, but emptied himself, taking the form of a servant" *Philippians 2:4–7, ASV.*

PRAYER

Lord Jesus, teach us to love as You loved. Help us to love our neighbor as ourselves, and set us free from all forms of coveting. Amen.

LESSON 4

Trusting God and Being Content

I am a firm believer in the truth that the best defense is a good offense. I believe if we focus on the positives we can ward off the negatives. Instead of examining ourselves regularly to see whether we are coveting, I suggest we take the opposite tack and pray for and love our neighbor. The character trait that a Christian who is free from coveting possesses is contentment. Following is a compendium of scriptures with this thought in mind.

"Be ye free from the love of money; content with such things as ye have: for himself has said, I will in no wise fail thee, neither will I in any wise forsake thee" *Hebrews 13:5, ASV.*

"But godliness with contentment is great gain: for we brought nothing into the world, for neither can we carry anything out; but having food and covering we shall be therewith content. But they that are minded to be rich fall into a temptation and a snare and many foolish and hurtful lusts, such as drown men in destruction and perdition" *1 Timothy 6:6–9, ASV.*

"Give me neither poverty nor riches; feed me with the food that is needful for me: lest I be full, and deny thee, and say, Who is the Jehovah? Or lest I be poor, and steal, and use profanely the name of my God" *Proverbs 30:8b–9, ASV.*

"Better is little with the fear of the Lord than great treasure and trouble therewith" *Proverbs 15:16.*

"Better is a little, with righteousness, than great revenues with injustice" *Proverbs 16:8, ASV.*

The Christian alternative to coveting is being thankful and content. Focusing on what God has provided helps you have a grateful heart. Contentment is an effective antidote to the wiles of advertising, which foster discontent and focus on what you don't have and usually don't need. In the eyes of the world, your car is always too old, your house is too small, your furniture is worn out, your computer is a dinosaur, and you deserve a vacation in paradise. We could go on and on along this line of thinking, and I am sure you could add a few examples yourself. You get the idea. The prime objective of advertising is to get your eyes on what you don't have, instead of what you do have. There is nothing inherently wrong with advertising. It is simply communicating the goods and services that are available. If the accountant that I have now had not advertised, I would not have known he was available or that he lived nearby. However, the media has taken advertising a step further in seeking to fuel discontent and subtly foster the notion that things will make you happy, instead of God.

As we remember the great hymn and "count our blessings, naming them one by one," we will "discover what the Lord has done." It is a learned skill to direct our hearts and minds in this manner, and it is one that Paul the apostle had mastered.

"Not that I speak in respect of want: for I have learned, in whatsoever state I am, therewith to be content" *Philippians 4:11.*

Contentment and a thankful heart are worth pursuing.

PRAYER

"Incline my heart unto thy testimonies, and not to covetousness. Turn away mine eyes from beholding vanity, and quicken me in thy ways" *Psalm 119:36–37, ASV.*

Help us to experience with Paul the secret of contentedness: "I know how to be abased, and I know also how to abound: in everything and in all things have I learned the secret both to be filled and to be hungry, both to abound and to be in want. I can do all things in him that strengtheneth me" *Philippians 4:12–13, ASV.* Amen.

Count Your Blessings

Johnson Oatman, Jr., 1897, music by Edwin O. Excell

When upon life's billows you are tempest tossed,
When you are discouraged, thinking all is lost,
Count your many blessings, name them one by one,
And it will surprise you what the Lord hath done.

Count your blessings, name them one by one,
Count your blessings, see what God hath done!
Count your blessings, name them one by one,
And it will surprise you what the Lord hath done.

LESSON 5

The Heart of a Man

The word *heart,* as used in scripture, does not refer to the organ responsible for distributing blood throughout the body. Neither is it to be interpreted as a passion or emotion. Rather, it is the very essence of who you are. It is the inner man. I believe the heart is the most important part of a Christian. Unless God transforms us in the innermost man, we have no hope. All of our external behaviors are just window dressing if we are still the same inside. It is for this reason that we receive Jesus by faith into our hearts.

"That Christ may dwell in your hearts through faith" *Ephesians 3:17, ESV.*

"Because if thou shalt confess with your mouth Jesus as Lord, and shalt believe in your heart that God raised him from the dead, thou shalt be saved: for with the heart man believeth unto righteousness; and with the mouth confession is made unto salvation" *Romans 10:9–10, ASV.*

There are around 862 references to the heart in scripture. Several verses follow to aid you in your study of the heart. David, one of the central figures in the Old Testament, is described as a man after God's own heart. He was chosen by God, through the prophet Samuel, because of his heart.

"But the Lord said unto Samuel, Look not on his countenance, or on the height of his stature; because I have rejected him: for the Jehovah seeth not as man seeth; for man looks on the outward appearance, but the Jehovah looketh on the heart" *1 Samuel 16:7 ASV.*

The new covenant given in Jeremiah 31 centers on the heart.

"But this shall be the covenant that I will make with the house of Israel; After those days, saith the Lord, I will put my law in their inward parts, and write it in their hearts; and will be their God, and they shall be my people" *Jeremiah 31:33.*

It is because of the importance of the heart that money, and our attitude towards it, is so important. Where and how we spend the money given to us by God provides a window into the condition of our hearts. Scripture says that the heart is deceitful and difficult to fathom.

"The heart is deceitful above all things, and desperately wicked: who can know it?" *Jeremiah 17:9.*

We can't check into a hospital and look into our spiritual hearts like doctors do with our physical hearts. But since we know that where our treasure is there also is our heart, we can use this truth to our advantage.

"For where your treasure is, there will your heart be also" *Luke 12:34.*

I heard an experienced financial counselor say that he could tell people's priorities by looking at their checkbooks. The same could be said about their credit card statements. Where we are spending and investing our treasure reflects our deepest convictions. A financial statement that has a record of every penny we spent in the course of a year would provide an excellent reading of the condition, direction, and priorities of our inner man. How we spend, invest, and think about money provides a glimpse into our innermost being. It is an indicator of the condition of our hearts. If you recognize habits or tendencies to view your material possessions in a way that is not godly, I suggest you run to God and ask Him for a heart transplant.

"A new heart also will I give you, and a new spirit will I put within you; and I will take away the stony heart out of your flesh, and I will give you a heart of flesh. And I will put my Spirit within you, and cause you to walk in my statutes, and ye shall keep mine ordinances, and do them" *Ezekiel 36:26–27, ASV.*

"Create in me a clean heart, O God; and renew a right spirit within me" *Psalm 51:10.*

When the heart is right, the wallet will be right. The converse is also valid. If your wallet is experiencing problems, take heed and ask God to show you what is not right in your heart. Ask Him to search, heal, and change your heart.

PRAYER

"Search me, O God, and know my heart: try me, and know my thoughts, and see if there be any wicked way in me, and lead me in the way everlasting" *Psalm 139:23–24.* Amen.

ADDITIONAL SCRIPTURE

"But he is a Jew who is one inwardly; and circumcision is that of the heart, in the spirit not in the letter; whose praise is not of men, but of God" *Romans 2:29, ASV.*

LESSON 6

Health, Wealth, and Prosperity

There is unmistakable teaching in scripture that God sometimes blesses His people with material things. Consider Job in the 42nd chapter of the book that bears his name.

"So the Lord blessed the latter end of Job more than his beginning: for he had fourteen thousand sheep, and six thousand camels, and a thousand yoke of oxen, and a thousand she asses" *Job 42:12.*

For those who do have wealth and have been blessed in this way, it is wise to consider: "Thou shalt remember the Lord thy God, for it is He that giveth thee power to get wealth" *Deuteronomy 8:18.*

When we do receive blessings from the hand of God, we are to be grateful and rejoice, for "every good gift and every perfect gift is from above, coming down from the Father of lights" *James 1:17, ASV.*

"Charge them that are rich in this present world, that they be not highminded, nor have their hope set on the uncertainty of riches, but on God, who giveth us richly all things to enjoy" *1 Timothy 6:17, ASV.*

REMEMBER

But never mistake a lack of material "blessings" as some kind of indication that God is not blessing you or doesn't care for you. That is clearly not taught in scripture. While Abraham, Job, David, and Solomon had significant possessions, Jesus had nowhere to lay His head, and the beloved disciple John lived out his days in exile in Patmos. There have been those who didn't receive the promises as material rewards, yet had just as much faith as those who did. God had some better things in store.

"They were stoned, they were sawn asunder, they were tempted, they were slain with the sword: they went about in sheepskins, in goatskins; being destitute, afflicted, ill-treated (of whom the world was not worthy), wandering in deserts and mountains and caves, and the holes of the earth. And these all, having had witness borne to them through their faith, received not the promise, God having provided some better thing" *Hebrews 11:37–40, ASV.*

THE DOWNSIDE OF WEALTH

Scripture includes other accounts of men who had a problem with their wealth, like the rich young ruler and the man who was intent on building bigger barns. He "layeth up treasure for himself, and is not rich toward God" *Luke 12:21.*

God knows what we can handle. Perhaps if we did have an abundance of possessions and riches, we would be tempted to forsake God. "They that will be rich fall into temptation and a snare, and into many foolish and hurtful lusts, which drown men in destruction and perdition" *1 Timothy 6:9.*

It is also easier to forget God when you have plenty to eat and a "goodly" home. This is spelled out clearly in *Deuteronomy 8:11–14:* "Beware that thou forget not the Lord thy God, in not keeping His commandments, and His judgments, and His statutes, which I command thee this day: lest when thou hast eaten and art full, and hast built goodly houses, and dwelt therein; and when thy herds and thy flocks multiply, and thy silver and thy gold is multiplied, and all that thou hast is multiplied; then thine heart be lifted up, and thou forget the Lord thy God, which brought thee forth out of the land of Egypt, from the house of bondage."

Consider the parable of the sower in Mark 4, Matthew 13, and Luke 8. "And others are the ones sown among thorns. They are those who hear the word, but the cares of the world and the deceitfulness of riches and the desires for other things enter in and choke the word, and it proves unfruitful" *Mark 4:18-19, ESV.*

God's word is eternal and universal. The truth being taught in the parable of the sower applies to all men, at all times. It seems to me this parable is particularly applicable to our affluent western culture. We run after possessions and things, and seem blind to our materialistic tendencies. This is the deceitfulness of riches. We are tricked into thinking that more money equals more happiness. And we are deceived. We work harder and put in longer hours and even encourage our spouse to provide a second income in this vain and fruitless quest for riches and the mirage of happiness. And as a result, the cares of the world and the deceitfulness of riches and the desires for other things enter in and choke the word, and it proves unfruitful.

Consider also the exhortation: "For what is a man profited, if he gain the whole world, and lose or forfeit his own self?" *Luke 9:25, ASV.*

THE UPSIDE OF WEALTH

"Charge them that are rich in this world . . . that they do good, that they be rich in good works, ready to distribute, willing to communicate; laying up in store for themselves a good foundation against the time to come, that they may lay hold on eternal life" *1Timothy 6:17–19.*

If God has given you material blessings, use those gifts as ministry opportunities to help others and meet their needs.

THE PROPER BALANCE

Recognizing the propensity of the human heart, and the need for proper balance, the inspired writer of Proverbs put it this way:

"Remove far from me vanity and lies: give me neither poverty nor riches; feed me with food convenient for me: lest I be full, and deny thee, and say, Who is the Lord? or lest I be poor, and steal, and take the name of my God in vain" *Proverbs 30:8–9.*

If you do receive material blessings, rejoice, and see that you are rich towards God. But if you don't, rejoice anyway. For God still loves you and has promised to never leave nor forsake you.

"Be ye free from the love of money; content with such things as ye have: for himself hath said, I will in no wise fail thee, neither will I in any wise forsake thee" *Hebrews 13:5, ASV.* Notice that the context of this scripture is in regard to God's provision.

PRAYER

In this lesson, my prayer for each of you is "Beloved, I pray that in all things thou mayest prosper and be in health, even as thy soul prospereth" *3 John 1:2, ASV.* Amen.

LESSON 7

Honor the Lord with Your Substance

"Honour the Lord with thy substance, and with the first-fruits of all thine increase: so shall thy barns be filled with plenty, and thy presses shall burst out with new wine" *Proverbs 3:9–10.*

"Will a man rob God? yet ye rob me. But ye say, Wherein have we robbed thee? In tithes and offerings" *Malachi 3:8, ASV.*

I recognize that there is no specific command to tithe in the New Testament. But neither does it specifically mention "Thou shalt not steal" (*Exodus 20:15)*, and that command is still valid. My personal conviction is that tithing is still valid as well. Giving 10 percent to God upon being paid or receiving remuneration for any kind of work is a good practice. Since it all comes from God, I am glad to return 10 percent to Him. I don't like to say I "pay" tithes since it is His money already. Saying that I am "returning" it to God helps me think properly about giving and about God's ultimate ownership of all we are and all that we have.

David expresses this truth in his prayer in 1 Chronicles 29. It is a long passage but worth reading.

"Then the people rejoiced because they had given willingly, for with a whole heart they had offered freely to the Lord. David the king also rejoiced greatly. Therefore David blessed the Lord in the presence of all the assembly. And David said: 'Blessed are you, O Lord, the God of Israel our father, forever and ever. Yours, O Lord, is the greatness and the power and the glory and the victory and the majesty, for all that is in the heavens and in the earth is yours. Yours is the kingdom, O Lord, and you are exalted as head above all. Both riches and honor come from you, and you rule over all. In your hand are power and might, and in your hand it is to make great and to give strength to all. And now we thank you, our God, and praise your glorious name.

"'But who am I, and what is my people, that we should be able thus to offer willingly? For all things come from you, and of your own have we given you. For we are strangers before you and sojourners, as all our fathers were. Our days on the earth are like a shadow, and there is no abiding. O Lord our God, all this abundance that we have provided for building you a house for your holy name comes from your hand and is all your own' " *1 Chronicles 29:9–16, ESV.*

Generally I give or return tithes to the church where I am attending and where I am a member. This has been referred to as a "storehouse tithe" *Malachi 3:10.* This money is used to pay for the upkeep of the church property and the support of the personnel. I also like to write the 10 percent check as soon as I have been paid. It is a small thing, but it helps maintain the priority of putting God first in all things. There is a reference to giving quickly in Exodus:

"Thou shalt not delay to offer of thy harvest, and of the outflow of thy presses" *Exodus 22:29, ASV.*

Malachi 3 speaks of both tithes and offerings. I also plan on giving to other Christian works and classify this as offerings. I try to budget these and give regularly to missionaries, pro-life ministries, needy individuals and families, radio ministries, and other worthy groups that God has brought across my path. I usually talk these over with my wife, so we are in agreement on whom to contribute to and how much.

My Hebrew professor noted that in Malachi 3:10, the text should be more accurately translated "will meet every need" instead of "that there shall not be room enough to receive it."

"Bring ye the whole tithe into the store-house, that there may be food in my house, and prove me now herewith, saith the Jehovah of hosts, if I will not open you the windows of heaven, and pour you out a blessing, *that will meet every need*" *Malachi 3:10 ASV.* This seems consistent with other scriptures about seeking first the kingdom and having our needs cared for, as spoken of in the Sermon on the Mount.

"But seek ye first his kingdom, and his righteousness; and all these things shall be added unto you" *Matthew 6:33, ASV.*

There is another facet to the promise in Malachi 3:10. God says "prove me." He challenges us to put Him to the test and see if He will back up His written word. If *we* are diligent in returning the tithe to God, *He* will be faithful in seeing that our every need is met.

After almost 30 years of applying this principle, I can testify that God has provided for all of our needs. But notice the distinction between needs and wants. During my years in the ministry, I have seen people spend money unwisely and then wonder why God wasn't meeting their needs. My opinion is that it wasn't a lack on God's part but undisciplined and impulse spending on their part. God's Word is true. As we honor Him with our substance, He will take care of His people. Indeed, He will and does meet every need.

I recently read the *Wealth Conundrum,* by Ralph Doudera. In this excellent little volume he makes the following comments about tithing: "Tithing is one of God's tests to see if we put Him first. It sets in motion a law of reciprocity that brings prosperity.* It is a basic fundamental law with spiritual consequences. Tithing from the first fruits of your increase requires an element of faith, since it wouldn't require any faith if you paid tithes only if you had some funds left over after spending the rest."

PRAYER

Help us, Lord, to honor You with our tithes and offerings and forever put You first in all things. Amen.

* I don't believe this promise guarantees prosperity, but rather that God will see that all of our needs are met when we do tithe.

LESSON 8

Glorifying God

This lesson is difficult to write because I am still learning so much about it myself. I confess that I have been guilty of making an unscriptural distinction between the secular and the spiritual. In my reading recently, I have seen it suggested that this dichotomy is much more prevalent in western culture than in other parts of the world. I don't know whether that is true or not, but I know that I still deal with this mindset and am trusting God to renew my mind along this line. This dichotomy is revealed when describing a person's occupation. Some will say they are working for the Lord if they are in church work or a form of missionary outreach. Those who are providing a living for their families and are not in a religious field will say they are working for themselves. Others will say they are involved in part-time or full-time Christian work. Those in the full-time Christian occupations are thus perceived as building treasure in heaven while others are working for filthy lucre and treasure on earth. This way of thinking and viewing our work is not biblical. When Christians do what God has called them to do, all of their labors are for God and His glory.

"Whether therefore ye eat, or drink, or whatsoever ye do, do all to the glory of God" *1 Corinthians 10:31, ASV.*

Certainly, there are those who are called to be pastors, teachers, and evangelists. Others are called to be pilots, doctors, and carpenters. When you are living for Jesus, all that you do is for God's glory. It doesn't matter what you are doing, as long as you are doing it unto the Lord. In the divine economy, work was created by God and is therefore good. Even before his fall, Adam was still given the task of tending the garden. Yes, "sweat" and "toil" are added to work as a result of the curse, but work is good (more on this in another lesson). May God set us free from the lie that some jobs are secular and others are sacred.

In order to learn this lesson, I have lived and worked in both camps. When I rededicated my life to Christ in college, my first thought was that I wanted to serve Him with everything in me. So I dropped one of my majors, took religion classes, and went to seminary. After graduation, I entered the ministry as an assistant to a pastor and as a youth leader. Subsequently I was ordained as the pastor. We were living in a church home where I was preaching, teaching, and overseeing summer youth camps. In those years, I believed I was serving the Lord 100 percent. During that same time period of my life, I also worked as a part-time cabinet maker, painter, and school teacher.

In the providence of God, we went through some difficult valleys, and I left the "full-time" ministry to teach at a Christian school and get our family back together. I began selling educational products and ended up writing a math curriculum. So now I am a businessman. I am still preaching, teaching, and speaking at conventions, and I am president of a corporation with several employees. Am I serving God any less? Absolutely not. My devotion to God has never changed. I am still serving the Lord with all of my time, talents, gifts, and energies. The only difference is that now I receive a check for my salary from my business instead of from the church treasury. By God's grace, I am still working for the glory of God. In hindsight, I see that God used these experiences to begin balancing out my tendency to distinguish between what is sacred and what is secular.

Having been on both sides of the coin has given me a unique perspective, for which I am grateful. I was unbalanced. I am becoming more balanced and as a result more scriptural. Work is good. In my business, I have also learned that when you work, you are contributing to the greater good of the larger community. Every business is a part of a team. I have excellent people that I work with, from the printers, box manufacturers, and plastic-injection molders on the production end to my office staff and sales reps on the marketing end, to United Parcel Service and United States Postal Service on the shipping end. I have been blessed by the talent and dedication of the people around me. I am glad they are working well to the best of their abilities and not wishing they were doing something more "spiritual."

On a personal note, I am grateful there were dedicated surgeons who had continued to do research and develop their skills when my eight-month-old baby needed open-heart surgery years ago.

Bottom line, it is not what you do but whom you do it for! I choose to work for the King of kings. Whether I eat or write math books or teach God's Word, I am doing it to the glory of God.

PRAYER

Help us to have the mind of Christ as we view our labor. Enable us to live and work unto the Lord. And whether eating or drinking, inspire us to do all to the glory of God. Amen.

LESSON 9

Compensating Church Work

As I stated in the previous lesson, my wife and I did work "for the church" for several years. I know that there are many different thoughts about how church workers, missionaries, and other full-time Christian workers should be paid. But as in every area of life, let's see what the Bible has to say about compensating those who do receive their wages from a church, free-will offerings, or a mission agency. Consider the following verses:

"In all labor there is profit" *Proverbs 14:23, ASV.*

"Thou shalt not muzzle the ox when he treadeth out the grain" *Deuteronomy 25:4, ASV.*

"The labourer is worthy of his hire" *Luke 10:7.*

It is okay to receive a legitimate wage, even if you are a full-time Christian worker. There is nothing in scripture to suggest that a servant of God should make less than everyone in his parish. If you are led to work for the church or a Christian ministry, do your best, but don't sell your talents or efforts short. You should expect to receive a livable wage. How much that amounts to in dollars and cents is up to those in your ministry, factoring in input from you and your spouse.

When discussing your compensation, communication up front is critical. If possible, get any proposals in writing.

If a portion of your salary package is living quarters, respectfully but firmly decline. When a church owns a parsonage or home that it gives you in exchange for a significant part of your income, it wins and you lose. The church has an appreciating asset, and if you leave in five years it still owns the home, which is increasing in value. You, on the other hand, have just lost five years of investing towards your retirement. For most people, their home, or estate, is what they will sell to provide income for their retirement years. My advice for those who are offered housing in lieu of salary is thanks, but no thanks.

1 Corinthians 9:1–14 is worth reading when considering the salary needs of a Christian worker. I have chosen to print only a few portions of this passage.

"What soldier ever serveth at his own charges? who planteth a vineyard, and eateth not the fruit thereof? Or who feedeth a flock, and eateth not of the milk of the flock? Do I speak these things after the manner of men? or saith not the law also the same? For it is written in the law of Moses, Thou shalt not muzzle the ox when he treadeth out the corn. Is it for the oxen that God careth, or saith he it assuredly for our sake? . . . If we sowed unto you spiritual things, is it a great matter if we shall reap your carnal things? . . . Even so did the Lord ordain that they that proclaim the gospel should live of the gospel" *1 Corinthians 9:7–14, ASV.*

Another scripture that should be considered is found in Galatians.

"So then, as we have opportunity, let us work that which is good toward all men, and especially toward them that are of the household of the faith" *Galatians 6:10, ASV.*

Whether you are a supportee or a supporter, endeavor to support those who are serving in the church and in other ministries as you would want to be supported if you were in their position. And remember that word *especially*: it says "especially" toward them who are of the household of faith. Even though I know that I need to beware of charlatans, those wolves in sheep's clothing who abuse these verses, my experience has been mostly on the other end of the spectrum, with those who have not been cared for in an honorable fashion. I have known devoted, faithful workers who have been overworked and underpaid. So despite a few who are abusing their positions, let's not forget to do good to those who are laboring faithfully in God's vineyard.

PRAYER

Once again we ask You to help us to think biblically about this issue. While we know there have been those who have used their positions in the ministry for their own gain, help us to not neglect those who through faith and patience are inheriting the promises. Amen and amen.

LESSON 10

Work is a God Thing

"Six days you shall labor, and do all your work, but the seventh day is a Sabbath to the Lord your God" *Exodus 20:9–10, ESV.*

Work is good. God called man to be a steward and tend the garden He created. Tending a garden is work. It is the task of a steward. And since Adam was given this task before the fall, we can assume man was created to labor, and this labor was good. After man fell, his occupation was coupled with sweat, and it became "toil." In our culture, work has not been viewed in a positive light, but originally it was good. Sin added the sweat of the brow and made it more like the drudgery that it is perceived to be today.

When we turn to Jesus and our sins are taken away, our lives are changed and we are restored to the proper relationship that He intended from the beginning. Can this transformation that makes us new creatures extend to our vocation as well? I think so. As Paul enjoins us, "Work hard and cheerfully at whatever you do as though you were working for the Lord rather than for people" *Colossians 3:23* (paraphrase). Our work is redeemed and ennobled as we labor for Jesus, joyfully tending His garden once again.

In discussing this subject as a family, we came up with some more applications. Work is loving your neighbor as yourself. If I love you and am painting your house, I will want to do a good job, just as I would want done for my house.

As the owner of *Math-U-See*, I think of this application in several different ways. One of my main goals is helping children understand mathematics. It is also my goal to help parents understand math and give them the tools and knowledge they need to help their children. If math becomes a positive experience, then the quality of life will be improved in their family. To that end, I am continually working to improve the product.

I also have sales representatives who have the same commitment to serve. I refer to them as math missionaries. They are servants at heart who get tremendous satisfaction out of helping people. They are also paid for their labor, which they should be. (This is a good thing for their homes!) I also have coworkers at our headquarters who take care of shipping, producing the books, ordering inventory, maintaining the web site, and doing other tasks. They, too, enjoy their jobs, receive a salary, and have the satisfaction of knowing they are helping people. Then there are those who assemble our block sets and fraction kits, put labels on the videos, build the wooden boxes, and put together DVDs. They are part of this team, as are the companies that print our books, duplicate our DVDs, film our videos, ship our materials, and heat our warehouse. All these jobs are required to help children understand math. If each member of this cooperative effort is working well and doing his job to the best of his abilities, it is a rewarding, fulfilling, and positive experience. This is good work; this is not drudgery.

My wife and I also do a lot of work around our home. That is another part of the garden we are tending. Personally, I didn't always like to keep things looking nice, but at one time we lived with an elderly Christian couple who had deep-rooted convictions that the place where you live ought to be properly maintained and cared for, to the glory of God. Another man whom I had tremendous respect for felt the same way about caring for his vehicles. His cars were cleaned and maintained regularly, inside and out. I learned this attitude toward physical possessions from these older, godly Christians. Sandi and I have worked to incorporate these attitudes and make them part of our lives. We see ourselves as stewards of our property and our vehicles. We have also found that when our physical surroundings are pleasant, they contribute to the emotional/spiritual part of us. It is rejuvenating and uplifting to walk through a well-kept garden, or visit a well-designed and well-ordered home. Nice things, tastefully arranged, are pleasing and add to the overall atmosphere of a home.

As we discussed this lesson in our family worship time, I found that my personality is such that I get a sense of satisfaction from making improvements and accomplishing tasks. My wife, on the other hand, just plain enjoys organizing, cleaning, and bringing order to our home. We enjoy having people come to our home for visits and meals. Having the physical environs decorated well contributes to the experience of those we are hosting. Many of our guests have commented on this in our guest book. Once again, our work is for others and is a part of serving them.

On another tack, as a man, I am keenly aware of how much my own sense of self-esteem is tied to my work and my success in the workplace. It is rewarding to know I am making a positive contribution to the lives of others. This knowledge makes my job satisfying. So while it is a good thing to help others, work also helps me. Being diligent and honest are not my natural characteristics. They had to be learned. Part of being a disciple is developing new and godly character qualities. Work is a great place to foster integrity and godly character. One of the great things about God is His ability to redeem people and situations. He takes work, even after the fall, and redeems it and uses it to build character and create opportunities for us to serve one another!

I hope these few paragraphs have helped you to have a new appreciation for tending or working in God's garden! Work is good because God is *good*.

PRAYER

Father, thank You for creating work for us to do. Help us to see this topic from Your perspective and joyfully embrace what You have designed. Amen.

LESSON 11

Providing for Your Family

"But if any provideth not for his own, and specially his own household, he hath denied the faith, and is worse than an unbeliever" *1 Timothy 5:8, ASV.*

The context of this scripture is caring for widows. See verses 3 and 16: "Honor widows that are widows indeed. . . . If any woman that believeth hath widows, let her relieve them, and let not the church be burdened; that it may relieve them that are widows indeed" *1 Timothy 5:3, 16, ASV.*

So while verse eight is speaking directly to the situation with widows, it also sums up what is obvious: the responsibility of parents to provide for the physical well-being of their family. It is a strong scripture and a powerful admonition. Providing for my "own household" is my job as a man and as a Christian. Not only is the man enjoined to provide for his family, but the worthy woman is also described as one who "looks well to the ways of her household."

"She riseth also while it is yet night, and giveth food to her household, and their task to her maidens. . . . She is not afraid of the snow for her household; For all her household are clothed with scarlet. . . . She looketh well to the ways of her household, and eateth not the bread of idleness" *Proverbs 31:15, 21, 27, ESV.*

It is my strong conviction that God created the man to be the priest and provider of the home. And while both parents are called to take care of the needs of the household, the responsibility of earning an income to provide for the needs of the family generally falls on the shoulders of the husband. I know of some unique circumstances where that is not the case, but those are not normal, in my opinion. Being a responsible Christian man includes being a diligent provider of food, clothing, and shelter. Perhaps I could restate 1 Timothy 5:8 to emphasize the positive contribution of the provider, instead of comparing him to an unbeliever who doesn't provide well.

If any provides for his own, especially his own household, he has affirmed the faith and is a true believer.

Caring for the basic necessities of your family is practical Christianity. It is a hallmark and characteristic of a true believer and is the legitimate fruit of a life devoted to Jesus. A father's labor in diligently providing for his family is just as "spiritual" as reading the Bible with his wife and children. To the wife and kids, may I suggest something? If you have a roof over your head, food on your table, and clothes on your back, find your dad or husband and say thank you.

"Render to all their dues: tribute to whom tribute is due; custom to whom custom; fear to whom fear; honor to whom honor" *Romans 13:7, ASV.*

PRAYER

Give us this day our daily bread and make us ever mindful of our primary responsibilities to our families. Amen.

LESSON 12

Seek First the Kingdom

LIVING BY FAITH

There is certainly a balance between being a diligent provider and living by faith. These are not mutually exclusive. Unfortunately, it took me a while to learn this lesson. Before I was a Christian, if I needed money, I worked for it. After I rededicated my life to serving the King, I began reading books about godly men and women, like Hudson Taylor and George Müller. From them I began learning about the life of faith, or praying for your needs to be met. These men told God about their needs, without telling any people. Only God knew what they needed. They trusted Him to communicate their needs to other people. As we know, except for manna from heaven, most of our needs are met through people on earth and not angels from heaven. There is plenty of scripture to support this idea, which is where these early faith missionaries got the idea. Hudson Taylor and George Müller were led by the Spirit to communicate their needs to God alone and not to any earthly individuals.

"But seek ye first his kingdom, and his righteousness; and all these things [food, clothing] shall be added unto you" *Matthew 6:33, ASV.* (The full text is *Matthew 6:25–33.*)

"And He called unto Him His twelve disciples. . . . Get you no gold, nor silver, nor brass in your purses; no wallet for your journey, neither two coats, nor shoes, nor staff: for the laborer is worthy of his food" *Matthew 10:1, 9–10, ASV.* (See also Mark 6:7–8 and Luke 10:4.)

It isn't wrong to communicate your needs to people. Paul spoke of the collection for the poor in Romans 15:26, Galatians 2:10, and 2 Corinthians 8 and 9. He was making a need known to the churches in his letters.

"For it hath been the good pleasure of Macedonia and Achaia to make a certain contribution for the poor among the saints that are at Jerusalem" *Romans 15:26, ASV.*

"Only they would that we should remember the poor; which very thing I was also zealous to do" *Galatians 2:10, ASV.*

As with most aspects of truth, there is a balance. During my first year of seminary, I had been working in the cafeteria a few hours per week as well as doing side jobs that would arise from time to time. But after I began reading about men of faith, I decided to let God do the providing. So I began saying no to the side work on the weekends and praying instead for God to meet my needs. I surmised that I was now "living by faith." In retrospect, I can see that part of me was sincerely trying to trust God and be "spiritual," but there was another part of me that was just plain lazy! I was expecting God to do all of the work. But as you will read in another lesson, I hadn't learned about meeting God halfway yet. It is interesting to note that during this time I got plain hungry, as funds were at an all-time low. I began to learn that I needed to continue to pray, yet not limit God to providing for my needs with a check in the mail or by cash appearing mysteriously in an envelope. God can provide supernatural manna from heaven if He chooses. He has that ability. During this period of my life, God provided work as an answer to my prayers, but I didn't recognize it. I have come to the conclusion that God expects us to be diligent in praying *and* working.

In 1983, God called our family to step out in faith and trust Him alone to provide for our needs. We lived that way for four years and never went hungry. Our supplications were directed to God and God alone. We had no visible means of support, but our invisible faithful God provided wonderfully. I also worked when the opportunity arose, as a part-time sign painter and a substitute teacher at the local high school. But this book wouldn't be complete or accurate if I didn't say that we have seen that God's hand has not been shortened over the years. If He calls you to live by faith, He will provide in the twenty-first century just as He did in the wilderness for the children of Israel, and in the first century for His disciples. The God of Israel, and the God of the apostles, is just the same today.

"For I the Lord do not change" *Malachi 3:6, ESV.*

"Jesus Christ is the same yesterday and to-day, yea and for ever" *Hebrews 13:8, ASV.*

PRAYER

Teach us how to look to You for all of our needs, and teach us the balance of living diligently by faith. Amen.

If you want to read more about the men of faith mentioned in this lesson, you can order these books or find more info at www.christianbook.com.

Janet Benge. *Hudson Taylor: Deep in the Heart of China.* Seattle: YWAM Publishing, 1998.

George Müller. *Autobiography of George Müller.* New Kensington: Whitaker House Publishers, 1984.

A.T. Peterson. *George Müller: All Things Are Possible.* Greenville: Ambassador International, 1999.

A.T. Pierson. *George Müller of Bristol.* Grand Rapids: Kregel Publications, 2000.

Howard and Mary Taylor. *Hudson Taylor's Spiritual Secret.* Chicago: Moody Press, 2009.

Hudson Taylor. *J Hudson Taylor*, Men and Women of Faith Series. Grand Rapids: Bethany House Publishers, 1987.

LESSON 13

Leaving an Inheritance

"A good man leaveth an inheritance to his children's children: and the wealth of the sinner is laid up for the just" *Proverbs 13:22.*

"House and riches are the inheritance of fathers, and a prudent wife is from the Lord" *Proverbs 19:14.*

Two of the reasons I work are to provide for my family now, and to provide for them for the future. Making money to pay the ongoing bills for my family has been covered in a previous lesson. In this lesson, I am going to address saving money for the future on two levels: the physical and the spiritual, although it is difficult to separate them. As a 54 year-old man, I have three future needs I am saving for: 1) emergencies that may arise; 2) our retirement; and 3) setting aside nest eggs for each of my sons.

Before I develop this thought any further, let me state very clearly that I have no confidence in the government to provide for me or mine. And I counsel you not to count on the federal government to give you your daily bread. Trusting in God and being diligent to lay aside money for the future is what I believe scripture teaches.

"It is better to trust in the Lord than to put confidence in man" *Psalm 118:8.* One way to do this is to set aside money from each and every paycheck. I have a friend who, upon getting his paycheck, sets apart 10 percent for God's work and another 10 percent for emergencies. He has done this for years. I confess that I was never that disciplined, especially when we had very little coming in and were living hand-to-mouth from paycheck to paycheck. We stayed in the black (no debts, thank the Lord) but didn't have much beyond that. Now that I have a regular salary and our house has been paid for, I have turned my attention to the three goals previously mentioned.

1. Emergencies—There are different schools of thought on how much you should set aside for needs that arise unexpectedly. Some say two months' salary, and others say six months'. Work this out for your family and then be intentional about setting aside money for this fund. You might open a savings account or invest in a money market fund. But put that money there and forget about it. It is for emergencies, not a new boat.

2. Our Retirement—If I were to pass away soon, I look upon our house, which is our personal estate, as a source of income for my wife and our special needs son. I also took out a term life policy for a 10-year period. Term life insurance is like renting insurance. If I were to die, my wife would receive the value of the policy. This way, my other three sons have a few years to get their financial feet under them. Then when they are more financially viable, they can contribute to the care of their mother and brother. This 10-year window covers their time spent in college and the first few years after their graduation.

3. Nest Egg—I do have a hope that my sons will not have to spend their best years working to get out of debt. People from my generation didn't own their own homes until they had lived in them for 30 years to pay off their 30-year mortgages. By that time, their grown children were married and repeating the same lifestyle, working off their own 30-year mortgages. By God's grace, my wife and I have a better hope and plan. We are living and saving so our kids will not have to sing and live "I owe, I owe, it's off to work I go." I would like to work with them so they are working for the next generation instead of a bank. Our hope is that if we stay out of debt, drive used cars, and avoid careless spending, we can help them get started in houses of their own without large mortgages. Then when their homes are paid for, they will be set free to begin setting aside a nest egg for each of their children. Instead of paying on a mortgage and trying to get out of a hole, they can begin investing in the future of their kids. Thus our generation, with God's help, will be the cycle breakers.

In addition to a physical inheritance, there is another viable inheritance that I am also seeking to leave, and that is loving God with all my heart, soul, mind, and strength. When I pass on to the next life, I hope my sons' primary memory of me will be of my love and devotion to the King and His kingdom. This is the ultimate inheritance, one that fades not away, reserved in heaven for you. (See 1 Peter 1:4.)

"Praise ye the Lord. Blessed is the man that feareth the Lord, that delighteth greatly in his commandments. His seed shall be mighty upon earth: the generation of the upright shall be blessed. Wealth and riches shall be in his house: and his righteousness endureth for ever" *Psalm 112:1–3.*

"Blessed be the God and Father of our Lord Jesus Christ, which according to his abundant mercy hath begotten us again unto a lively hope by the resurrection of Jesus Christ from the dead, to an inheritance incorruptible, and undefiled, and that fadeth not away, reserved in heaven for you" *1 Peter 1:3–4.*

There is a definite balance to be found here, as is true with most things. On one hand, I am preparing an inheritance for my sons as if the Lord may tarry in His returning. But whether He returns this year or in 20 years, I am seeking to leave them a vital, living faith in Jesus Christ. If He were to return today, my hope is that they would be taken up in the clouds to live and reign with Him forever.

This balance between the eternal and the earthly was evident in the inheritance given to the Levites. On the one hand, God was their eternal portion (Deuteronomy 10:9 and Numbers 18:20) instead of a portion of the land of Israel, which Levi's brothers received. On the other hand, they did receive physical cities as their earthly possession. (See Numbers 35:2.)

Ultimately, our top priority is heaven. But in that pursuit, let us also be mindful of our earthly responsibilities, which are also mentioned in scripture.

In my further study on this topic, I have observed that God has something to say on how an inheritance is distributed after it has been accumulated.

"An inheritance may be gotten hastily at the beginning; but the end thereof shall not be blessed" *Proverbs 20:21.*

As I read this verse, my thoughts went immediately to Luke 15 and the prodigal son.

"And he said, A certain man had two sons: and the younger of them said to his father, Father, give me the portion of goods that falleth to me. And he divided unto them his living. And not many days after the younger son gathered all together, and took his journey into a far country, and there wasted his substance with riotous living" *Luke 15:11–13.*

The prodigal son was certainly not blessed by receiving his inheritance as he did, "hastily at the beginning." It seems wise to specify in your living trust not only how much each of your heirs is to receive, but also how and when it is to be distributed.

PRAYER

Let us not be so heavenly minded that we are no earthly good. Rather, let us be so God minded that our families are cared for, now and forever. Amen.

LESSON 14

Marriage and Money

It seems that the majority of problems a married couple will face are related to money. Conflicts arise about who makes it, who spends it, how it is spent, whether they should go into debt, and what are their long-term goals concerning retirement, etc.

I would hope that most of these questions would be discussed at length and in depth *before* you get married. Doing so allows you to be pulling in the same direction. Some will say such deliberations are not "spiritual" (I may have said that when I was 25, *sigh*), but recall the scripture on the front of this book.

"For where your treasure is, there will your heart be also" *Luke 12:34.*

In other words, when you discuss the topic of money in a relationship, you are really getting down to the nitty-gritty of who you are. Your heart-to-heart discussions will reveal the life priorities you each have. It is the hidden man of the heart (see 1 Peter 3:4) that ultimately determines our life choices. So endeavor to cover the topic of money thoroughly with your potential spouse.

Any discussion about finances and marriage will also impact other areas of the Christian life. As the next few paragraphs deal with the husband/wife relationship, it is inevitable that we will also touch on the concept of Christian leadership and headship. In my brief tenure on this earth, of which 26-plus years have been spent as a Christian husband and 20 have involved different aspects of pastoral ministry, I have come to the following conclusions about 1 Peter 3:7.

"Likewise, ye husbands, dwell with them according to knowledge, giving honour unto the wife, as unto the weaker vessel, and as being heirs together of the grace of life; that your prayers be not hindered" *1 Peter 3:7.*

In our relationship, my wife and I have come to the understanding that it is my job to earn the income and pay the bills. I am the provider. God has made me to carry responsibility and bear burdens differently than my wife. Ultimately, Jesus bears all of our burdens, but being responsible for the physical needs of a home is a manly function. And when a man accepts this responsibility, it contributes to the making of him. It is part of the training that enables a man to be manly. I have learned to love shouldering the responsibility for my family's needs. It is my niche. It is not always easy, but it is good for me in particular and for men in general to bear this burden. My helpmeet could probably do this job, and do it well, as she does all of her other jobs. But she has chosen to let me be the man, and in doing so, I have let her be the woman.

Here is an example of how this division of labor has played out in our home. When we were living from paycheck to paycheck, which we did for many years, I slept well. If Sandi had known how tight our finances were, she would not have slept well. It took me several years to learn this about her, but that is just the way she is made up. I know personality is a factor here as well, but I have talked this idea out at seminars for years and have heard many, many testimonies that verify it. A man is made to be the priest and provider of the home. As 1 Peter 3 says, our wives are the weaker vessel and we are to dwell with them accordingly. I could give many examples of this, but I am going to stay within my own observations and experiences.

As I have said, this posture not only protects and nurtures the femininity of the wife, but it also encourages and fosters the masculinity of the husband. Men are being criticized for not being masculine any longer. I don't know how to rectify the situation completely, but the area of finance is a huge piece of the puzzle. The very idea of a husband going hat in hand to his wife and asking for his weekly allowance is emasculating. Let men be the provider of the home and responsible for the disbursement of the finances, and it goes a long way toward enabling him to be who God created him to be.

Another aspect of this concept has to do with the wife as queen of the castle. Since my wife and I are joint heirs of the grace of life, we have consulted together and come up with a budget for the running of our "castle," which is my wife's department. Each month, a portion of our income is deposited into the household account. She then allocates the money to the different areas within her sphere, including food, clothing, vitamins, and household supplies. Obviously, there are times when needs arise that are not a part of the everyday workings of the home, and they have to be addressed, but this doesn't happen as much as you would think. As Sandi is caring for these needs, it is my responsibility to pay the heating bill, electric bill, insurance costs, and car expenses, as well as tithes, etc. We have been using this method for years and it has worked famously. I don't need to be involved with every nickel in the day-to-day running of the castle, and she is free to budget it, with the help that God gives to a queen.

Large-ticket items, such as carpet, drapes, and furniture, come out of the general budget, and I write the checks for those, even though they are within the sphere of the household. Because these purchases will impact everyone in the house, she usually asks for input from all of us. But after giving our two cents worth, we defer to Sandi to make the final call since that is her jurisdiction. This method is played out in a similar fashion with items in my jurisdiction, such as buying vehicles. We discuss and pray over these together before moving ahead.

THE HOUSEHOLD BUDGET

There are many helpful tomes written on the subject of personal budgeting. I recommend anything by Crown Ministries, which you can find at www.crown.org. But as I mentioned in a previous unit, my personal philosophy has been to spend less than I earn, contribute to the work of God, and set aside what I can for the future and emergencies. My wife, on the other hand, is a disciplined enveloper and was trained this way at an early age. When I say *enveloper*, I think you know that I am referring to having literal white envelopes for specific expenses and future planned purchases. Each of us has to find what works for us.

For those of you who will be married one day, this lesson is not about your own personal budgeting but household budgeting. You are going to have to work through a budget with your spouse. A discussion about priorities and budgeting can be a special time of getting your hearts together. It may also be potentially explosive. Remember, when you are discussing money, you may be pushing buttons that are at the very essence or heart core of each other's being. So pray, make sure you are rested, and set aside a specific time when you can give this time of communication your full attention. Some couples may need a third party to walk them through this sensitive arena.

Budgeting is common sense applied to your finances. The first step in developing a budget is ascertaining how much money you have and how much you will continue to have coming in from your paycheck. This is your income, or in-come. If you have a salary, this is easy to predict. But if you are self-employed, this amount will vary from month to month and week to week. So take several weeks and months to figure out your average income.

Then you will need to determine how much will be going out, or your out-go. These are your living expenses. The way we arrived at our home budget was by a process of tracking what we spent over a period of time. I can't overemphasize the importance of doing this. Take a month, or preferably several months, and keep track of what you and your wife spend and then tabulate your results into categories, such as food, clothing, utilities, books, recreation, gas, car expenses, and phone bills. Once you have concrete data, which will vary from year to year, you can begin to make some informed decisions about how to spend your money in a way that is honorable before God.

On a practical note, I have found that using a computer program for personal finances is efficient and helpful. With a touch of a button, you will know how much of your money is being spent and where it is going. I would also suggest carrying a note card in your wallet for chronicling cash outlays on a daily basis. Then, at the close of the day, transfer your expenditures to a ledger or notebook at home. With the computer tracking your credit card and checking, and your ledger carrying the cash disbursals, you now have reliable data to help you know your expenses, or what is out-going. After you receive a couple of paychecks, you will know what is in-coming. With this data, you are well on your way to developing a working budget for your home.

PRAYER

God, help us to submit ourselves one to another in the fear of God and so live as to glorify You in the day-to-day handling of Your resources. Thank You for supplying funds for our daily bread. Give us the discernment necessary to determine our income as well as our outgo, and grant us grace to have the discipline necessary to live within what You have provided. Amen.

"Subjecting yourselves one to another in the fear of Christ" *Ephesians 5:21, ASV.*

LESSON 15

I Owe, I Owe, It's Off to Work I Go

Debt is a spiritual issue. My main problem with those who are perpetually spending beyond their means, with their out-go surpassing their in-come, is that they presume upon the future.

"Come now, ye that say, To-day or to-morrow we will go into this city, and spend a year there, and trade, and get gain: whereas ye know not what shall be on the morrow. What is your life? For ye are a vapor, that appeareth for a little time, and then vanisheth away. For that ye ought to say, If the Lord will, we shall both live, and do this or that" *James 4:13–15, ASV.*

"Do not boast about tomorrow, for you do not know what a day may bring forth" *Proverbs 27:1, NASB.*

We have no assurance of the future. If we accumulate personal debts, we have no assurance that we will have the funds to pay those debts. The number of personal bankruptcies that occur every year is staggering. A few weeks out of work because of an accident or illness, and many people will never recover financially. For example, a worker experiences a job layoff for a few months, and because he is so dependent on his weekly paycheck, he will not be able to make his car payment for a month. Just check the local paper to see how many cars are being repossessed because of this.

The economy will go through downturns, and modern technology will continue to replace jobs. History tells us these things will happen. I am not trying to create fear, just simply stating that we don't know the future, which is why God says we are to literally say, "Lord willing," when speaking of our hopes and plans. Only God knows what is to come, and it is prudent to let Him lead us by the principles in His Word. It is also wise to commit your life and affairs into His care regularly.

Having debt is not sin, but it is usually not wise. The children of Israel were not forbidden to borrow money. In their culture debts were canceled in the seventh year.

"At the end of every seven years you shall grant a release. And this is the manner of the release: every creditor shall release what he has lent to his neighbor. He shall not exact it of his neighbor, his brother, because the Lord's release has been proclaimed" *Deuteronomy 15:1–2, ESV.*

So while borrowing money is not a sin, take care that you are not presumptuous. There may be occasions when you borrow money, but do it in the fear of God and not in a spirit of presumption.

"Keep back thy servant also from presumptuous sins; let them not have dominion over me: then shall I be upright, and I shall be clear from great transgression" *Psalm 19:13, ASV.*

When it comes to borrowing money for a house or for a car, there are a few factors to consider. If you are considering purchasing a house or land, recognize that historically they tend to appreciate, which means their value increases with time. This is not always the case, but generally it is. When we lived in Massachusetts, there was an incredible growth in property values. But the growth did not continue and there was a market correction. Some people who were in the process of buying and selling when the market regained a sense of sanity lost their shirts. Others who sold at the right time in the growth spurt made a tremendous profit. At the beginning of the twenty-first century, many baby boomers seem to be taking their retirement funds out of the stock market, which has been volatile of late, and investing in real estate. This influx of money is driving real estate values to record highs, but they will eventually return to normal levels, in my opinion, so be careful.

Other items, particularly cars, depreciate and lose their value over time. That is why I am much more open to borrowing money for a home than a new car. There is more on this topic in the instruction manual where we will compare dollars and cents.

If you are being led to purchase a home and can borrow money from a Christian brother or close relative, it seems preferable to going in debt to a bank. Whichever avenue you believe God is leading you to choose, make sure you get plenty of godly counsel and careful communication.

A good rule of thumb in any financial undertaking is to make sure you understand the costs involved. In the math section of this course, we will study the many costs associated with owning a home and operating a car. When you have an accurate idea of the costs involved and can compare them with the amount of resources you have, then you can make an informed decision and avoid going into debt in the first place.

"For which of you, desiring to build a tower, doeth not first sit down and count the cost, whether he have wherewith to complete it?" *Luke 14:28, ASV.*

The wisest and safest course is to not presume upon the future but to live within your means. God will provide for your needs. Trust Him.

PRAYER

"Keep back thy servant also from presumptuous sins; let them not have dominion over me: then shall I be upright, and I shall be innocent from the great transgression" *Psalm 19:13.* Amen.

LESSON 16

Giving

"One can give without loving, but one cannot love without giving."

– Amy Carmichael

"Every man according as he purposeth in his heart, so let him give; not grudgingly, or of necessity: for God loveth a cheerful giver" *2 Corinthians 9:7.*

Giving is the fruit of a redeemed life. The evidence of a heart transformation is an open wallet. Where your treasure is, there will your heart be. Conversely, where your heart is, there will your treasure be. I was told of a tight-fisted individual who was converted and baptized. He became a wonderful example of an open-handed, giving Christian. When asked what had made the difference, he remarked that when he was immersed he still had his wallet in his back pocket, and it was baptized as well. When God grants us a new heart, a new wallet will follow. As a new creature in Christ, you will find yourself thinking of others and their needs. This is the beginning of loving your neighbor as yourself.

"Wherefore if any man is in Christ, he is a new creature: the old things are passed away; behold, they are become new" *2 Corinthians 5:17, ASV.*

There are many ways to love your neighbor, and there are many needs besides financial needs, but since these lessons are geared toward how we spend God's money, we are going to focus on that aspect. Historically, giving without any expectation of return, which is a good definition of love, is a uniquely Christian virtue. In his book *Under the Influence*, Alvin J. Schmidt sets forth his thesis of how Christianity—including Christian giving—has transformed civilization.

On a practical note, my father-in-law used to say, "If you don't give when you have two dollars, you won't give when you have two hundred dollars." The January 2005 issue of *Money Matters*, published by Crown Financial Ministries, had a similar headline, "People who do not give philanthropically as youngsters are less likely to do so as they mature and age." So let me encourage you to make it a habit to give early and often, and then when God does provide you with more funds, you will have developed the habit of being a cheerful giver. I began giving as a young Christian, and there is a lot of truth in what the article says. Faithfulness in little things prepares us for bigger things.

"He that is faithful in a very little is faithful also in much" *Luke 16:10, ASV.*

"And he said unto him, Well done, thou good servant: because thou wast faithful in a very little, have thou authority over ten cities" *Luke 19:17 ASV.*

This could also be an explanation of the Bible story about the widow's mite.

"And he saw a certain poor widow casting in thither two mites. And he said, Of a truth I say unto you, This poor widow cast in more than they all" *Luke 21:2–3, ASV.* She certainly was rewarded, not for how much she gave but for giving what she could when she didn't have much.

When you give, do it without fanfare. That way God gets the glory (and no one will ask you for money!). When someone has been praying in secret, it is a wonderful thing to see God answer the same way. I have been on both ends of many of these situations. I have been the one praying who became a grateful recipient, and I have been the channel for God's answer and experienced the satisfaction of knowing I had worked with God.

"But when you do alms, let not your left hand know what your right hand does" *Matthew 6:3, AKJV.*

Recently, I learned that most nonprofit ministries receive 40 percent or more of their donations for the entire year in one month, December. I can think of two reasons for this. The first is that many self-employed individuals don't know how much they have made until they meet with their accountant at the end of the year. And secondly, it is the last chance to make a tax-deductible donation.

Let me encourage you to give regularly as a part of your monthly budget. This makes it easier for the ministries you are supporting to be effective. And since these ministries are a part of your regular budget, it is easier to explain to solicitors representing other worthwhile causes that all of your money set apart for giving has already been allocated.

Before you give to an organization, do your homework. Researching is much easier with the advent of the Internet, and I recommend two sources. The first is the EFCA,

Evangelical Council for Financial Accountability (www.efca.org); and the other is Ministry Watch (www.ministrywatch.com). Both of these sites will provide you with pertinent information about many of the nonprofit Christian ministries operating today.

Another lesson I have learned, and which we will develop more in another segment, is to pray before giving. Just because you are aware of a need doesn't mean you have to meet it yourself. There is a definite balance between having an open hand and being led by the Spirit. First of all, as a husband I have to remember that my first priority is to my family. When I used to sell books and education resources as our main source of income, it was very tempting to give books away to a needy family. But we were a needy family as well. I felt impressed by the Holy Spirit to check with my wife first before giving away resources that were taking away from my own responsibilities as provider.

Sometimes God just wants you to pray. It may be that He wants to provide in a different way. It is a good thing to be willing to meet someone's felt need. But since we don't want to get ahead of God's best for the individual that we are concerned about, it is a good idea to ask God first.

PRAYER

"If any of you lack wisdom, let him ask of God, that giveth to all men liberally, and upbraidth not; and it shall be given him" *James 1:5.*

Father, give us wisdom to know when to give, and how to give. Cause us to be led by Your Spirit in this crucial area. And most of all, transform our hearts as well as our wallets so we are cheerful cooperators with You in being faithful givers. Amen.

LESSON 17

The Gift of Giving

THE GRACE OF GIVING

"But first gave their own selves to the Lord, and unto us by the will of God. Insomuch that we desired Titus, that as he had begun, so he would also finish in you the same grace also. Therefore, as ye abound in every thing, in faith, and utterance, and knowledge, and in all diligence, and in your love to us, see that ye abound in *this grace* also" *2 Corinthians 8:5–7* (emphasis added).

In reading the eighth chapter of 2 Corinthians, I am struck by the repeated use of the word *grace*. Grace, we know, comes from God and is something He bestows. It is a gift. The Corinthians were being exhorted to receive a particular kind of grace, the grace to give. The fact that I can become a giver by asking for the grace to give is very encouraging to me. It is really good news! When I read these words, I realize that whether I am a giver by nature or not, I can become a giver by Christ's grace and His super-nature. Grace is received by asking and believing for it.

"But this I say, He that sows sparingly shall reap also sparingly; and he that soweth bountifully shall reap also bountifully. Let each man do according as he hath purposed in his heart: not grudgingly, or of necessity: for God loveth a cheerful giver. And God is able to make all grace abound unto you; that ye, having always all sufficiency in everything, may abound unto every good work" *2 Corinthians 9:6–8, ASV.*

I know it is God's will for me to be a cheerful giver. And since we know it is according to His will, we know we can expect to receive "this grace also." Let me encourage you to ask for, and receive, the grace of giving.

"If we ask anything according to his will, he heareth us: and if we know that he heareth us whatsoever we ask, we know that we have the petitions which we have asked of him" *1 John 5:14–15, ASV.*

GIVE AND IT SHALL BE GIVEN

A bonus of becoming a gracious giver is that you will reap what you sow. When you become a faithful giver, you can expect God to take care of your needs. "And my God shall supply every need of yours according to his riches in glory in Christ Jesus" *Philippians 4:19, ASV.*

Many times I have been in prayer services where a specific need was expressed. Then a well-meaning Christian would quote this verse in his prayer with the assumption that since God will supply all of our needs, then He will supply this one. It was years before I understood the context of this promise and realized it was probably being misquoted. It is a wonderful discipline to memorize scriptures such as Philippians 4:19, which is a popular one, but we know

that when reading scripture, the context is just as vital in understanding a particular verse as the passage itself. Let's consider the background of this promise. In 2 Corinthians 8, Paul is talking about giving. He is also speaking about the same issue in Philippians 4.

"And ye yourselves also know, ye Philippians, that in the beginning of the gospel, when I departed from Macedonia, no church had fellowship with me in the matter of giving and receiving but ye only" *Philippians 4:15, ASV.*

Clearly, the Philippians had received the grace of giving. They were a giving church. In fact, it says they were the only church to respond to the needs as outlined by Paul. So within this context, he is saying that since they had given so faithfully before, God would meet their needs now. Because they had given, God would give to them. Paul was undoubtedly familiar with the next two scriptures.

"Give, and it shall be given unto you; good measure, pressed down, shaken together, running over, shall they give into your bosom. For with what measure ye mete it shall be measured to you again" *Luke 6:38, ASV.*

"The liberal soul shall be made fat; and he that watereth-shall be watered also himself" *Proverbs 11:25, ASV.*

When you are a gracious giver, you will most certainly find that God is a gracious and faithful provider. As you faithfully give to others, you qualify for the fulfillment of Philippians 4:19. When you give to others, God shall supply every need of yours.

But recall how this lesson began: God is able to impart to us the grace to be givers. He will supernaturally see that we abound in "this grace also" according to His riches and glory in Christ Jesus

PRAYER

God, our faithful provider, we ask for "this grace also" to be channeled to us so that we would be givers, in word and in deed. Amen.

LESSON 18

Offerings

As I mentioned previously, tithes belong to God, and generally they are returned to the church where you are a member. Offerings are given in addition to tithes.

"Will a man rob God? yet ye rob me. But ye say, Wherein have we robbed thee? In tithes and offerings" *Malachi 3:8, ASV.*

But where and to whom do we give these portions? I suggest that there are five groups of people.

YOUR NEIGHBOR

The first folks are those around you. You don't have to look abroad to find needy families and individuals. Perhaps you have relatives who are in need. Scripture refers to them as your neighbors and your brothers.

"Withhold not good from them to whom it is due, when it is in the power of thy hand to do it. Say not unto thy neighbor, Go, and come again, and to-morrow I will give; when thou hast it by thee" *Proverbs 3:27–28, ASV.*

"But whoso hath the world's goods, and beholdeth his brother in need, and shutteth up his compassion from him, how doth the love of God abide in him?" *1 John 3:17, ASV.*

Keep your eyes and ears open to those close by. There have been times when I suspected someone needed a helping hand, and so I asked him if he did. Upon learning more about his circumstances, I prayed and felt led to give. Sometimes I have learned of needs in church when a family or an individual asked for prayer. I have also found out about other situations from my wife. It seems that she has her own connections and is often made aware of needs that I know nothing about. So first pray for the people, and then ask God if He wants you to be a part of the answer. The answer doesn't always have to be money. Offer to wash their car, take their kids for a trip to the park, or mow their lawn. There are many practical ways to help, and all of them will be a blessing. God will lead you. I always wanted to make sure I wasn't saying, "Be warmed and be fed" and then turning my back and walking away. That would not be loving my neighbor. "If a brother or sister is poorly clothed and lacking in daily food, and one of you says to them, 'Go in peace, be warmed and filled,' without giving them the things needed for the body, what good is that?" *James 2:15–16, ESV.*

THE WIDOWS AND THE FATHERLESS

"Pure religion and undefiled before our God and Father is this, to visit the fatherless and widows in their affliction" *James 1:27, ASV.*

Widows and orphans are always on God's heart and are safe to support at any time. I look at single mothers as widows. Many single moms have been left by men who are not supporting them, and they are worthy of any help you can give them. Their children might also qualify as "fatherless." Taking them under your wing and mentoring them might be the best thing you can do for them.

THE POOR

Jesus said the poor are always with you, but you still can support them. Handouts are not always the best way. A growing portion of our society has grown up as part of the welfare generation. Unfortunately, they are trained to think they are entitled to a free ride. A gift to someone with this mindset is usually counterproductive. A job and a chance to work will produce better results that will improve his character and self-esteem in ways that a handout never could.

Knowing whom to give to can be difficult and requires discernment. You want to be responsive, but you also don't want to get ahead of God. He may have other means of meeting someone's needs. When working with God, offer yourself, be willing to do whatever, and then pray and trust God to lead you.

More scriptures for your edification:

"He that despiseeth his neighbor sinneth; but he that hath pity on the poor, happy is he" *Proverbs 14:21, ASV.*

"He that oppresseth the poor reproacheth his Maker; but he that hath mercy on the needy honoreth Him" *Proverbs 14:31, ASV.*

"Jehovah will root up the house of the proud; but he will establish the border of the widow" *Proverbs 15:25, ASV.*

"He is gracious to a poor man who lends to Lord and he will repay him for his good deed" *Proverbs 19:17.*

"Whoso stoppeth his ears at the cry of the poor, He also shall cry, but shall not be heard" *Proverbs 21:13, ASV.*

"There is that coveteth greedily all the day long; but the righteous giveth and withholdth not" *Proverbs 21:26, ASV.*

"Rob not the poor, because he is poor; neither oppress the afflicted in the gate: for Jehovah will plead their cause " *Proverbs 22:22–23.*

"Remove not the ancient landmark; and enter not into the fields of the fatherless: for their Redeemer is strong; He will plead their cause against thee" *Proverbs 23:10–11, ASV.*

"He that giveth unto the poor shall not lack: but he that hideth his eyes shall have many a curse" *Proverbs 28:27.*

"The righteous taketh knowledge of the cause of the poor; the wicked hath not understanding to know it" *Proverbs 29:7, ASV.*

"If there be with thee a poor man, one of thy brethren, within any of thy gates in thy land which the Lord thy God giveth thee, thou shalt not harden thy heart, nor shut thy hand from thy poor brother; but thou shalt surely open thy hand unto him, and shalt surely lend him sufficient for his need in that which he wanteth" *Deuteronomy 15:7–8, ASV.*

"For the poor will never cease out of the land: therefore I command thee, saying, Thou shalt surely open thy hand unto thy brother, to thy needy, and to thy poor, in thy land" *Deuteronomy 15:11, ASV.*

Two more parts of the body of Christ worthy of support are full-time Christian workers at home and abroad. Your pastor, a summer camp ministry, or missionaries laboring in a third-world culture all come under these two headings. When I read of Paul in Acts, I associate him with a foreign missionary taking seriously the great commission in Matthew 28:18–20.

CHRISTIAN WORKERS AT HOME

When I read the Old Testament and come across verses that apply to the Levites, I think of my pastor. In Nehemiah, the people were rebuked for not caring for the Levites and forcing them to abandon the work of God. They were forced to work in the fields to meet their daily needs.

"And I perceived that the portions of the Levites had not been given them; so that the Levites and the singers, that did the work, were fled every one to his field" *Nehemiah 13:10, ASV.*

If a person is called to preach the gospel, he should be recompensed accordingly. Unfortunately, many Christian workers are underpaid. Consider giving in addition to their salaries.

"For the scripture saith, Thou shalt not muzzle the ox when he treadeth out the corn. And, The laborer is worthy of his hire" *1 Timothy 5:18, ASV.*

Doing unto others as you would have them do to you is always the best attitude to have when supporting individuals. In a few years, you will have a better understanding of what it takes to support a family. When my family and I lived in the church parsonage in Georgia and I served as pastor, there was one couple who gave us a significant check every year that we used for buying Christmas gifts. We didn't have the money for Christmas presents, and that was a huge blessing that Sandi and I still recall to this day. It was intensely practical and met a specific need that we had.

CHRISTIAN WORKERS ABROAD

Then there are your brothers and sisters in Christ who are laboring in different countries and cultures. These men and women need our support. Missionaries are particularly difficult to support regularly since they are often out of sight and therefore out of mind. I hope you have the opportunity to serve God abroad at some point in your life, whether for three months or three years. Not only will you get to do good, but doing so will also open your eyes to the needs of the world and the opportunities to serve God abroad as nothing else can.

"Go ye therefore, and make disciples of all the nations" *Matthew 28:19, ASV.*

If you have attended a Christian college, some of your fellow students will no doubt be led to serve God in this manner. Keep in touch with them and support them as they do. Or perhaps you will have a chance through your local congregation. Attending and learning at mission conferences is also a part of preaching the gospel to the ends of the earth.

The added benefit of supporting missionaries and others is that when you invest in their ministry, your heart is drawn to them in prayer. I have prayed a lot more for India because of a children's home in Kerala, India, which we supported. The same is true for many of the missionaries we have have supported over the years. Put your treasure there and your heart will be more inclined to follow. I learned this from my mother-in-law, who certainly had the gift of giving and supported many worthy causes for over 40 years.

"For where thy treasure is, there will thy heart be also" *Matthew 6:21, ASV.*

If you are led to support missionaries in Europe or some other industrialized country, recognize that their needs are going to be greater than those of an indigenous worker in the hill country of India. The cost of living is so different. Expect their needs to be similar to your own in the U.S.

Most individuals serving the Lord in full-time ministry will tell you their greatest need is prayer support. Giving should increase your supplications on their behalf, but watch that it doesn't replace them. Give *and* pray.

"And he said unto them, The harvest indeed is plenteous, but the laborers are few: pray ye therefore . . ." *Luke 10:2, ASV.*

PRAYER

Father in heaven, help us to work with You in caring for the poor and needy. Give us discernment as to where and how to give in a way that truly loves our neighbor as our self. And thank You for the privilege of supporting Your servants at home and abroad. Bless us so that we might bless others in Jesus' name. Amen.

LESSON 19

Purchasing Begins with Prayer

When making a major purchase, such as a car, a home, or furniture, there are four concepts that have helped me in making wise and godly decisions. I will deal with these in four lessons. The first concept is prayer.

PRAY

"Be careful for nothing; but in every thing by prayer and supplication with thanksgiving let your requests be made known unto God" *Philippians 4:6.*

Notice the exhortation to pray in all things. Prayer involves God in decisions. It is amazing to me the difference it makes when you pray about everything. It has been standard practice in our home to pray when we leave the driveway, as we "hit the road," whether the trip is long or short. I can remember many times, especially at the beginning of long trips, what would happen to our memories as we were praying. It became a joke. As we prayed, we began to think of things we forgot to bring, or people we had forgotten to contact, or appliances we had forgotten to turn off. Once we began to pray, it seemed that God entered our thinking and directed our thoughts to these different things. It has happened so many times that I usually begin praying before we leave the driveway.

When you ask God to help, He will help. He is a gentleman and appreciates being asked. He waits to be invited and does not barge in without an invitation.

"Ye have not, because ye ask not" *James 4:2.*

"Ask, and it shall be given you" *Matthew 7:7.*

In everything, pray and ask, even when the answer seems obvious. In Joshua there is an illustration of this principle. Joshua and the children of Israel were experiencing God's blessing on their campaign to inherit the land of Canaan. The Gibeonites, who lived nearby, were afraid to meet them in battle, so they devised a scheme to preserve their lives. They came to Joshua dressed as if from a long journey and negotiated a deal with him and the other leaders. Joshua and the Israelite leaders fell for the ruse completely. The whole situation is summed up in verse 14.

"And the men took of their victuals, and asked not counsel at the mouth of the Lord" *Joshua 9:14.*

It seemed so obvious what to do, they thought they didn't need to ask counsel of God. Had they prayed, and had they asked, I believe God would have helped them. Had they put their energies into trusting God instead of their own brainpower, they wouldn't have made this costly mistake.

"It is better to trust in the Lord than to put confidence in man" *Psalm 118:8.*

God likes to reveal Himself and help His children. All we need to do is ask in faith.

"For the eyes of the Lord run to and fro throughout the whole earth, to show Himself strong in the behalf of them whose heart is perfect toward Him" *2 Chronicles 16:9.*

We are encouraged to ask God for our daily bread.

"Give us this day our daily bread" *Matthew 6:11.*

I take this to mean not just bread, but our everyday needs as well. After over 30 years of serving Him, I am happy to report that He does provide for our daily needs. David saw it, and so have I.

"I have been young, and now am old; yet have I not seen the righteous forsaken, nor his seed begging bread" *Psalm 37:25.*

One way that He has provided for us is by giving us a little ingenuity and initiative after we pray. We have not been averse to shopping at thrift stores and consignment shops. We have frequently gone to yard sales (garage sales in some parts of the U.S.) and kept our eyes open to ads in the local paper, penny saver, or advertiser. We have seen remarkable answers to specific prayers, and it is good for our pride to do a little scrounging. Humility and a dependence on God are always good for the Christian. When you pray, keep your eyes open for the answer. It may come in unexpected ways and places.

PRAYER

Father, help us to trust in the Lord with all our heart and lean not on our own understanding. In all our ways cause us to acknowledge You, for we know You shall direct our paths (adapted from *Proverbs 3:5–6*). Amen.

LESSON 20

Purchasing with Discernment

DISCERNMENT

"Ye ask, and receive not, because ye ask amiss, that ye may consume it upon your lusts" *James 4:3.*

"Walk in the Spirit, and ye shall not fulfil the lust of the flesh" *Galatians 5:16.*

When you read these scriptures, notice the word *lust*. Let me suggest a different word that conveys the same thought but makes it more up to date. The word is *desire*, or *strong desire*.

"Ye . . . receive not . . . that ye may consume it upon your **desires**" *James 4:3.*

"Walk in the Spirit, and ye shall not fulfill the **strong desire** of the flesh" *Galatians 5:16.*

After praying (principle 1), ask God to help you discern whether the item you want to purchase is a need or a strong desire. If you haven't figured it out by now, we each have a tremendous potential to rationalize any action or purchase. This is part of our nature, our flesh, which is represented by a strong desire. Not all desires are wrong. God often gives us desires that are good. So don't fall into the trap of thinking that if it feels good, it must be wrong. That is why we need discernment.

Discerning the difference between needs and desires takes time and maturity. Here are some tips to help you with this process.

COUNSEL

First, talk with your spouse. She knows you and will help you to see things more clearly.

"Submitting yourselves one to another in the fear of God" *Ephesians 5:21.*

Communicate to your wife, the best you can, all of the pros and cons. It may help you to write them down. Then discuss it with your parents and close friends. Remember, this is not something you need to do when deciding whether you need new shoes; this is for major decisions like whether to borrow money to buy a car or a house. "Where no counsel is, the people fall: but in the multitude of counsellors there is safety" *Proverbs 11:14.*

Here is a recent example of how I applied these principles to a significant purchase. In the spring of 2004, I was in need of a riding lawn mower to cut our field across the creek.

We had an old one-hundred-dollar special that was erratic, slow, and required frequent repairs. I checked into hiring a lawn mowing service. If I hired them, it would cost $30.00 each time they came. Since I need it cut from April to November, or eight months, at four times per month, that is 32 cuts. So to hire it out would cost 32 x $30.00 or $960.00.

I spoke with my helpmeet about the lawn mower situation, and she agreed that we needed to make a change. Then I bounced the idea off a few friends whose insights I respected.

The local hardware store carries a popular name-brand mower with all the bells and whistles. I didn't need all the extras, since the area I was cutting was more of a field than a tricky lawn with a lot of corners, trees, etc. The name-brand mower retailed for about $1,500.00. I found another one at a department store that was a reliable model but not nearly as flashy for $800.00. There was pressure to buy the flashy name-brand model, but it wasn't what I needed. Discernment helps you determine whether you really need a flashy mower or whether you desire a name brand to impress your friends and neighbors. That is another reason to proceed carefully, to sort out what is the will of God and what is pressure from the world. We purchased the mower for $800.00 and it has worked well for us.

There is no one answer to every purchasing question. There are many factors to consider. If I were using the mower for the front yard, instead of the level field, it might have been a good idea to get the one with the fancy transmission and the tight turning radius. But for our needs at that juncture, we saved $700.00.

So pray for wisdom and discernment in deciding what are needs and what are desires; then discuss it with your spouse and some friends. After this process, you should be on your way to making a wise and careful decision.

PRAYER

God in heaven, save us from being swayed by our own desires, and help us to clearly identify what we need and when we need it. Amen.

LESSON 21

Purchasing Carefully

You have heard the saying "Haste makes waste." Here is the scriptural basis for that idiom.

"He that believeth shall not make haste" *Isaiah 28:16.*

Salesmen understand the need to close a sale quickly after their presentations. When your mind is full of all the positives they have trotted out, they know that the best time to seal the deal and get your signature on the dotted line is right away. But a night's sleep and some careful thought go a long way towards restoring sanity about their proposals. I can give many examples of slick proposals and polished sales pitches.

Our first experience occurred when we were young newlyweds and were invited to see some lakefront property and receive a free gift for doing so. We wanted the "free" gift, and it seemed like a fun outing, so along we went (like sheep to the slaughter). We listened to a pitch about buying a lot (with yearly maintenance fees) around a man-made lake an hour south of our church home. It was a measly sized piece of land, thick with young scrub oak trees and the size of a large campsite. But, there was going to be a clubhouse and a beach, and we could "almost see" the lake, since the lakefront properties were already sold. We could own our own plot of land for a mere $6,000.00. We actually thought about it, and may have signed on, except I knew the principle of not being in haste. Much to the sales staff's dismay, we said we wanted to think about it overnight. They began throwing in added incentives to get us to sign up that day. Our faces were set like flint, and the more they saw we were really leaving, the more determined they were to have us purchase then. Pretty soon the head salesman was standing there giving us his best shot. Fortunately, we escaped to the safety of our home. The more we thought about the whole setup, the sillier we felt. First of all, we rarely got away for a weekend, since we were serving in a church. Secondly, we seldom went camping, and when we did, preferred to go to different places, not the same one each time. Thirdly, it was an hour away. And lastly, believe it or not, 25 years ago in that area of Georgia, you could buy a whole acre of land for $1,500.00 instead of a portion of an acre for $6,000.00. If we wanted to buy a piece of land for getting away, we could spend the same amount and find four acres somewhere closer.

TIMESHARES

I have heard several carefully crafted presentations about buying timeshares. A timeshare is a nice piece of property like a condominium or townhouse. If you buy one, you are buying the property for a share of the time. It may be a townhouse that is worth $100,000.00, and instead of buying it yourself, you buy it along with several other families.

Since you own it for only one week of the 52 in a year, you just pay for a portion of the total cost, say $8,000.00. Each year you are entitled to that vacation property for one week. In addition to your initial investment, you also incur an annual maintenance fee of several hundred dollars.

Without going into all the pluses and minuses, I want to stress again the pressure there is to buy the day you hear the spiel. I have never purchased a timeshare, but I know how different the proposition looks the next day. What always got me was the maintenance fee. The several hundred dollars one had to cough up annually could be used for a comfortable getaway by itself. Living in a church parsonage with very limited resources, we had learned how to make a dollar stretch. So for the same money as the maintenance fee, we could have had a nice vacation, and we didn't have to pay the cost of the initial investment. And have you ever noticed all of the timeshares listed in the paper and on eBay that are for sale? I know a family that couldn't sell their timeshare, so they opted to give it away for free. They ate their initial investment just to be rid of the rising maintenance fees. It was cheaper in the long run to write it off than to keep making those yearly payments.

The other side is that my parents have purchased several timeshares, and we have enjoyed many nice vacations with our extended family.

I could go on and tell you many stories about encyclopedia demonstrations, buying clubs, new cars, etc. When you pray and give it a day or so, they all look a lot different.

Now for a final balancing comment. If you have prayed, discerned a need, consulted with your spouse, and done your research, and you discover just what you need, buy it! I watched a certain car for over a year. When I saw exactly what I wanted on eBay I bid on it and won the bid. We got just what we were looking for and at a reduced price. When you move promptly to seal the deal, do so in a spirit of gratefulness, recognizing God has led you, and not in a spirit of being driven or pressured. Personality studies show that there are some people who aim, aim, aim . . . and never fire. And there are those who fire and then aim. So work to find the proper balance between researching and purchasing.

PRAYER

Thank You for Your Word that directs us to believe and not be in haste. Thank You for the peaceable fruit this has produced in our family so many, many times. Your Word is indeed a lamp to our feet and a light to our path. Amen.

LESSON 22

Purchasing with Peace

Being led by the Spirit of God is my last piece of purchasing wisdom. It is definitely connected to praying and asking God for assistance, but it is more about understanding how God directs us and speaks to our hearts. Being led is one of the many benefits of having a God who knows how to communicate with each one of us.

"For as many as are led by the Spirit of God, these are sons of God" *Romans 8:14, ASV.*

Often God speaks to us through another individual, and more often than not, through His written word. But there are times He communicates with us directly.

"The Spirit himself beareth witness with our spirit" *Romans 8:16, ASV.*

I am assuming, of course, that you have prayed, consulted with your spouse and godly friends, and have identified this purchase as a definite need. But there are times when we have done all we know to do, and it still isn't clear which way to turn or what decision to make. My initial observation is that if you are on the right track, you can expect a sense of peace. If you are on the wrong track, there may be a check in your spirit or a nagging feeling that won't go away. The presence of peace or the lack of peace are generally reliable indicators of whether the Prince of Peace is leading you.

"Her [wisdom's] ways are ways of pleasantness, and all her paths are peace" *Proverbs 3:17, ASV.*

"Now the Lord of peace himself give you peace at all times in all ways. The Lord be with you all" *2 Thessalonians 3:16, ASV.*

"And let the peace of Christ rule in your hearts, to the which also ye were called in one body; and be ye thankful" *Colossians 3:15, ASV.*

Learning to hear God's gentle Spirit and to be led by Him isn't an easy thing for me to describe, as it seems to be different for everyone. How God leads you is a part of the relationship you are developing with your heavenly Father. Christianity is certainly based on truth, but it is more than a formula; it is a living relationship with God. I am giving tips from my experience of walking with God, but you each must develop your own hold on God. When you consider scripture, you see all the men and women had unique experiences of being led. Abraham communicated with God directly. Moses heard Him speak from a burning bush. Joshua saw Him as the commander of the hosts of the Lord. I could go on and on citing examples of how differently God speaks to His children. God made us each unique, and He knows how to lead each one of us uniquely.

Notice the experience of Elijah on the mountain.

"And he said, Go forth, and stand upon the mount before the Lord. And, behold, the Lord passed by, and a great and strong wind rent the mountains, and brake in pieces the rocks before the Lord; but the Lord was not in the wind: and after the wind an earthquake; but the Lord was not in the earthquake: and after the earthquake a fire; but the Lord was not in the fire: and after the fire a still small voice. And it was so, when Elijah heard it, that he wrapped his face in his mantle, and went out, and stood in the entering in of the cave. And, behold, there came a voice unto him, and said, What doest thou here, Elijah? *1 Kings 19:11–13.*

God was not in any one of these spectacular phenomena. He was in the still small voice.

Another indicator is to make sure you are led and not driven. There should be no pressure to buy or not to buy, but rather a calm assurance that it is the right thing to do. It has been said that the devil drives, but the Lord leads.

"When he hath put forth all his own, he goeth before them, and the sheep follow him: for they know his voice" *John 10:4, ASV.*

I have raised goats and found them to be special animals. I have friends who have raised sheep and have nothing but unpleasant experiences to recount about the stupid animals. A pastor from Lebanon came to our church once, and I asked him why God identified His people as sheep instead of goats. He told us many stories, but in short, you lead a sheep but you have to drive a goat. Sheep will follow their shepherd, and they know his voice. To herd goats, the goat herder must be behind the flock with a handful of rocks and a stick. Be led, not driven or pressured, into making a decision. It is your money. Make a careful decision and spend it wisely.

PRAYER

Thank You for creating us and making us know that we are Your children. Lead us as we call upon You, and help us to recognize Your still small voice in our lives. Give us ears to hear, we pray. Amen.

LESSON 23

Just and Fair

God is just and fair. As He is, so are we to be in this world.

"Because as he is, even so are we in this world" *1 John 4:17, ASV.*

"A just weight and balance are the Lord's: all the weights of the bag are his work" *Proverbs 16:11.*

God appreciates fairness and justice in our dealings with one another, and He has very definite opinions about unfair business practices.

"Differing weights and differing measures, both of them are abominable to the Lord" *Proverbs 20:10, NASB.*

When I accepted Christ's offer of forgiveness of sins and the gift of eternal life, I received a new conscience. Things that I would have previously classified as insignificant were now important. When I went through a checkout line and received the wrong change, I found myself giving back what did not belong to me. I also found God leading me to go back to merchants whom I had stolen candy from as a kid. I not only asked for their forgiveness, but I also paid back what I had taken—with interest. This extended to athletic tape that I had taken while on the basketball team in college. This is called restitution.

It is important to be honorable in our dealings with others. It is really nothing tricky, just simply loving our neighbors as ourselves. After following Jesus for 30-plus years, I am grateful for the ways God has led me to be honorable in my dealings with customers and others with whom I do business.

"For we take thought for things honorable, not only in the sight of the Lord, but also in the sight of men" *2 Corinthians 8:21, ASV.*

"Render to no man evil for evil. Take thought for things honorable in the sight of all men" *Romans 12:17, ASV.*

Paying bills on time is honorable and is a tangible way to love your neighbor as yourself. As anyone in business will tell you, cash flow, or having enough money to operate with, is critical to not going into debt. If several people who owe money to a business decide to delay their payments, it often means the owner is unable to meet his obligations. Some large businesses have a reputation for delaying payment by three or four months instead of the customary 30 days. This is not honorable. I am amazed at how much positive feedback I have received for paying my bills in a timely fashion. Many of my suppliers have written letters and thanked me for paying promptly upon receiving their invoices. This is being honorable in the sight of all men, and God is glorified when His children behave this way.

If you hire someone who is poor to work for you, there are scriptures to guide you.

"Thou shalt not oppress an hired servant that is poor and needy, whether he be of thy brethren, or of thy strangers that are in thy land within thy gates: at his day thou shalt give him his hire, neither shall the sun go down upon it; for he is poor, and setteth his heart upon it: lest he cry against thee unto the Lord, and it be sin unto thee" *Deuteronomy 24:14–15.*

Pay them quickly when they have completed the tasks for which they were hired. They need the money right away, not 30 days hence.

PRAYER

Let our light so shine before men, that they may see our good works, and glorify our Father which is in heaven *(based on Matthew 5:16).* Amen.

LESSON 24

Getting Rich Quickly

Beware of "get-rich-quick" schemes. If they seem too good to be true, they probably are.

"He that maketh haste to be rich shall not be innocent" *Proverbs 28:20.*

"He that hasteth to be rich hath an evil eye, and considereth not that poverty shall come upon him" *Proverbs 28:22.*

In our day, gambling and lotteries seem to dominate the billboards and media advertising. No one benefits from them except the companies selling them. In fact, lotteries and other forms of gambling wouldn't make a profit if the thousands of losers didn't far outnumber the handful of winners. And if lotteries weren't profitable to the people sponsoring them, I guarantee there wouldn't be any. I know they market them to the general public as aiding the elderly and providing money to schools, but the few dollars thrown in the direction of "worthy causes" are far outweighed by the negative effect on society as a whole.

I have never purchased a lottery ticket. Neither do I intend to buy one. I like the story of the man who went through a checkout line in a supermarket where lottery tickets were on sale. He told the cashier that he won every time he played. She asked what his secret was. He told her he took the dollar in his hand and put it back into his wallet. He won every time!

There are those who say if they won the lottery, think of all the good they could do with the winnings. This is a classic case of rationalizing greed. It is an example of the end justifying the means. This kind of thinking is ludicrous and deserves no further comment.

But what about the "winners"? I have been told of some disturbing trends among lottery winners. It seems that when someone wins the lottery, he rarely has much to show for it a few years down the road, since people often spend their winnings frivolously and carelessly. Deep down they know they have not worked for the money and thus don't deserve the income. Then there are the pressures that instant wealth brings on marriages and friendships. More money is not the answer to life's problems—Jesus is. Coveting does not produce peaceable fruit, but godliness with contentment is great gain. On the *Math-U-See* website, there is an article published by the Associated Press that is a commentary on the effects of the lottery on a "winner." You may access it at www.forum.mathusee.com/lottery.

And what about the "losers"? It is commonly known that the majority of tickets are sold to people in lower income brackets who can't afford this costly habit. Not only do they invest money that should go to their families, but their character is affected as well. The lottery, like other forms of gambling, is based on getting something for nothing. This insidious lie has the potential to cause havoc in our homes today. We were not designed to get something for nothing. Man was created to work for his daily bread. Those who are pinning their hopes on instant riches do not put their hearts into legitimate labor but waste energy waiting for their miracle ship to come in. Work, the very thing that would provide for their families and dignify their existence, is thus discarded for the hope of striking it rich. All suffer, the bread winner and family alike. The family suffers from not having enough money for its daily bread, and the provider suffers in his innermost being, knowing he is not providing as he should. Much of a man's self-esteem comes from doing a good job in his place of employment. Being diligent and faithful in his daily work produces lasting fruit in his character and increased revenue in his pocket. A faithful and diligent provider will foster a healthy home and put bread on the table, clothes on the family members' backs, and a roof over their heads.

"Better is a little, with righteousness, than great revenues with injustice" *Proverbs 16:8, ASV.*

"A faithful man shall abound with blessings; but he that maketh haste to be rich shall not be unpunished" *Proverbs 28:20, ASV.*

"Wealth hastily gotten will dwindle, but he who gathers little by little will increase it" *Proverbs 13:11, RSV.*

"Wealth from get-rich-quick schemes quickly disappears; wealth from hard work grows over time" *Proverbs 13:11, NLT.*

"He that is greedy of gain troubleth his own house" *Proverbs 15:27.*

HOPE FOR ADDICTION

There are many ministries and agencies devoted to helping people overcome gambling addictions. I know that buying one lottery ticket does not signify an addiction, but if this is a problem for you or someone you know, get the help that is available. May God deliver us from addictions, open our eyes to the dangers inherent in them, and give us the grace to not give in to the temptation of hastening to be rich. I have been tempted to buy lottery tickets and participate in get-rich-quick schemes. And even if I had participated, I would still go to heaven. I would just be poorer. It isn't wrong to be tempted; Jesus was tempted just as we are. It is giving in to temptation that leads to sin.

"For we have not a high priest that cannot be touched with the feeling of our infirmities; but one that hath been in all points tempted like as we are, yet without sin" *Hebrews 4:15, ASV.*

"Wherefore also he is able to save to the uttermost them that draw near unto God through him, seeing he ever liveth to make intercession for them" *Hebrews 7:25, ASV.*

"Blessed is the man that endureth temptation; for when he hath been approved, he shall receive the crown of life, which the Lord promised to them that love him. Let no man say when he is tempted, I am tempted of God; for God cannot be tempted with evil, and he himself tempteth no man: but each man is tempted, when he is drawn away by his own lust, and enticed. Then the lust, when it hath conceived, beareth sin: and the sin, when it is fullgrown, bringeth forth death" *James 1:12–15, ASV.*

PRAYER

Once again, faithful Lord, help us to be content with such things as You have provided and to be ever mindful of Your Word and Your ways. Save us from looking to unprofitable means and methods to replace the tried and true labor of our hands which blesses the provider and his family. Amen.

LESSON 25

Standing Before Kings

In place of pursuing shortcuts to have our needs met, let's consider all of the scriptures in Proverbs that magnify diligence and warn against laziness.

"He becomes poor that deals with a slack hand: but the hand of the diligent makes rich. He that gathers in summer is a wise son: but he that sleeps in harvest is a son that causes shame" *Proverbs 10:4–5, AKJV.*

"The labor of the righteous tends to life: the fruit of the wicked to sin" *Proverbs 10:16, AKJV.*

"The hand of the diligent shall bear rule: but the slothful shall be under tribute" *Proverbs 12:24, AKJV.*

"He also that is slothful in his work is brother to him that is a great waster" *Proverbs 18:9, AKJV.*

"The sluggard will not plow by reason of the cold; therefore shall he beg in harvest, and have nothing" *Proverbs 20:4, AKJV.*

"Love not sleep, lest thou come to poverty; open thine eyes, and thou shalt be satisfied with bread" *Proverbs 20:13.*

"The thoughts of the diligent tend only to plenteousness; but of every one that is hasty only to want" *Proverbs 21:5.*

"Seest thou a man diligent in his business? he shall stand before kings; he shall not stand before mean men" *Proverbs 22:29.*

Diligence is meeting God halfway. In our home, we have often referred to the book and movie *Where the Red Fern Grows,* written by Wilson Rawls. It is centered around a young boy's desire to have some hunting dogs. In a moment of exasperation, the boy pours out his heart to his grandfather and tells of his countless prayers for coon dogs. The wise old man tells his grandson that perhaps God wants him to have more than dogs; He wants him to have character. Of course the boy tells him that he doesn't want character; he just wants dogs. But the grandfather then suggests that maybe he ought to meet God halfway by working (diligently) as well as praying. That is the point of diligence; it doesn't replace faith in God, it works with God. Genuine faith goes hand-in-hand with diligent work.

I learned this lesson just as the boy in the Ozarks did. When I was a young and zealous Christian, I read the biographies of several outstanding men, including Jim Elliot, Hudson Taylor, C. T. Studd, and George Müller. I thought you were not spiritual if you worked, but rather that you should pray and let God supply your needs. This seemed to be the high road. I saw work as unspiritual. Sadly, I didn't understand the balance of praying and working for many years. Later I learned that God can answer our requests in many different ways. Sometimes God gives us the money directly, and other times He gives us work to do to earn the money. Both are legitimate answers to prayer. Faith and works is a winning combination, as the book of James states so eloquently.

I place a high value on diligence and Christian character when I am looking for employees at my place of business. When I interview a contractor for anything from a small job around my home to a major construction project, the main thing I look for is not only the price, but the character of the worker. When I say character, I mean someone who is honest, fair, reliable, and diligent. I have dealt with many people in this vein, having served as a general contractor on two major renovations at our office building and for putting an addition onto our home. I also have several representatives who work with our company throughout the country and overseas. Over the years, certain traits have emerged in those who have been successful and certain characteristics have come to the fore in those who have had difficulty. It has very little to do with education or business expertise and everything to do with their character and their personal integrity. Business practices are extensions of the person. Diligent people stand out. Honesty is a hallmark. But then, I am not surprised, because that is what God's word predicts in the selections from Proverbs that we read at the beginning of this lesson.

Usually we have to learn diligence. Personally, I don't think I was as diligent as I should have been when I was just beginning to operate my own business, but God has His good ways of teaching us. One of them is hunger (smile).

"It is good for workers to have an appetite; an empty stomach drives them on" *Proverbs 16:26, NLT.*

A wise friend once shared with me her experiences helping budding entrepreneurs. Those who were hungry worked hard to make ends meet. In the process, they developed a positive work ethic that stayed with them even when they were doing well. Diligence had become a habit. On the other hand, those who were well financed when they were starting up their businesses didn't have to work as hard and often didn't succeed. Perhaps it was a lack of motivation (hunger), but for whatever reason, they didn't develop the same work ethic. When all else fails, go look at the insects.

"Go to the ant, thou sluggard; consider her ways, and be wise: which having no guide, overseer, or ruler, provideth her meat in the summer, and gathereth her food in the harvest. How long wilt thou sleep, O sluggard? When wilt thou arise out of thy sleep? Yet a little sleep, a little slumber, a little folding of the hands to sleep: so shall thy poverty come as one that travelleth, and thy want as an armed man" *Proverbs 6:6–11.*

"The ants are a people not strong, yet they prepare their meat in the summer" *Proverbs 30:25.*

PRAYER

Rabbi, make us diligent in the work that You give us to do. Amen.

"Take my yoke upon you, and learn of me; for I am meek and lowly in heart: and ye shall find rest unto your souls. For my yoke is easy, and my burden is light" *Matthew 11:29–30.*

LESSON 26

The Yoke's on You!

PARTNERSHIPS

As Christians, we are living for Christ. That is preeminent. When it comes to entering into a business partnership with an unbeliever, my advice is to avoid it at all costs. This alliance will not work. You are each living for different gods. A nonbeliever, regardless of his good intentions, is living for self. You, as a Christian, are living for Christ. Your business cannot serve two masters. You are like two oxen in one yoke pulling in different directions.

"Be ye not unequally yoked together with unbelievers: for what fellowship hath righteousness with unrighteousness? and what communion hath light with darkness? And what concord hath Christ with Belial? or what part hath he that believeth with an infidel? And what agreement hath the temple of God with idols? for ye are the temple of the living God; as God hath said, I will dwell in them, and walk in them; and I will be their God, and they shall be my people. Wherefore come out from among them, and be ye separate, saith the Lord, and touch not the unclean thing; and I will receive you, and will be a Father unto you. And ye shall be my sons and daughters, saith the Lord Almighty" *2 Corinthians 6:14–18.*

I would also suggest that you be very careful about entering into a partnership with a fellow believer. There are cases where it has worked, but it takes an extra dose of grace. Often relationships suffer, and the older I get, the more I value godly friendships. Such friendships are priceless. If you do feel that you should be a partner with another brother in Christ, apply all that I have said about prayer, counsel, etc., and go slowly. Then before you go any further, put all your expectations in writing. A contract may seem petty or less than trusting, but the clearer you are in the beginning about your expectations for one another—such as the ways you are each to be compensated and your individual responsibilities—the better off your relationship will be. A third party is helpful in putting your partnership in black and white and making sure all the bases have been covered. There are businesses that specialize in helping people set up just such a working agreement.

LENDING

When you lend money to an individual, you are also entering into a type of bond or agreement that has similar ramifications as a business arrangement. Most of the time we think of the bank as the institution that lends money, but sometimes parents lend funds to their children. My advice is to treat lending like a bank, in that you spell everything out in black and white and get both parties to sign on the dotted line. I am all in favor of Christians helping out other brothers and sisters in everything from small loans to financing the purchase of a home. I think we should be doing a lot more of this. But the more clearly you spell out the terms of repayment, the interest rate, how soon you can pay it off, etc., the better off all the concerned parties will be. Putting all of the expectations clearly on the table extends to parents and children as well. People all have different expectations, and the clearer you can be up front, the better off you will be in the long run. If there are older people involved in the deliberations—and there usually are since they have the most money—remember that they forget things. There is nothing dishonorable about a document; you are working to maintain and protect relationships. A lawyer or an accountant can be helpful in these situations.

"The rich ruleth over the poor, and the borrower is servant to the lender" *Proverbs 22:7.*

One of the reasons a relationship can become strained is because of this principle of borrower and lender. Regardless of how much money the lender has, it is *his* money. He or she probably worked hard to acquire it. If you do borrow from a parent or friend, be aware that doing so will introduce a new twist in your personal relationship. That person is now the lender and you are now the servant. This verse may never be mentioned, but the dynamic is now present, and always will be, as long as you owe someone money.

Jack Whittaker, winner of the 315-million-dollar Powerball in 2002, bemoans how relationships were affected when he says that nearly all of his former friends have asked to borrow money from him. In his own words he states, "And of course, once they borrow money from you, you can't be friends anymore." (You may access the entire article at www.forum.mathusee.com/lottery.)

On the upside, it is nice to know that a parent or friend holds your mortgage, not a bank. If an emergency arises, it is a lot easier to approach a merciful friend for an extension than the board of a lending institution, where you are only a statistic.

PRAYER

Save us from ungodly relationships. Help us to be wise in the matter of partnerships and the borrowing and lending of money. Cause us to be honorable in all of our dealings in Jesus' name. Amen.

LESSON 27

Guard Your Signature

COSIGNING AND SURETY

Another area that may affect you is cosigning on a note. When a friend or relative applying to a lending institution for a loan does not have any assets to put up for collateral, he will ask for someone else to cosign the loan. Now *collateral* and *assets* are simply fancy words for items of value. Let's say your friend wants to borrow $2,000.00. The bank wants to know what he has of value that it can have if he isn't able to pay back the loan, say a piece of property or a bag of gold coins. If the bank is going to lend him $2,000.00 in cash, it wants to be sure he has assets to take in the event that he stops making payments. This is common sense.

But if your friend does not have any assets, he will ask someone else to cosign the loan who does have the ability to repay it. A cosigning situation usually occurs when the person needing the money can not establish credit on his own, or doesn't have the necessary collateral, so he needs someone with good credit to cosign with him. Plain and simple: don't do it. Cosigning for someone else is foolish. There are reasons they can't get the credit on their own. More often than not, you will be left paying the debt and then have nothing to show for it but a painful lesson.

Proverbs is clear on this subject. If you really want to do it, be prepared for the worst case scenario: paying off the loan yourself.

"My son, if thou be surety for thy friend, if thou hast stricken thy hand with a stranger, thou art snared with the words of thy mouth, thou art taken with the words of thy mouth. Do this now, my son, and deliver thyself, when thou art come into the hand of thy friend; go, humble thyself, and make sure thy friend. Give not sleep to thine eyes, nor slumber to thine eyelids. Deliver thyself as a roe from the hand of the hunter, and as a bird from the hand of the fowler" *Proverbs 6:1–5.*

Here is a definition of surety from Noah Webster's 1828 dictionary:

Surety, *n.* - In law, one that is bound with and for another; one who enters into a bond or recognizance to answer for another's appearance in court, or for his payment of a debt or for the performance of some act, and who, in case of the principal debtor's failure, is compellable to pay the debt or damages.

PRAYER

"For freedom did Christ set us free: stand fast therefore, and be not entangled again in a yoke of bondage" *Galatians 5:1, ASV.* Amen.

WORKING WITH FRIENDS AND RELATIVES

When you are working on a project with close friends or relatives, be extra careful to put all of the expectations in writing. Explain that you have a certain amount of money for the project and you don't want to begin unless you have a specific quote. You may not think this is necessary, but the more time you spend clarifying all aspects of the work to be done, the better off all the parties will be.

"For which of you, desiring to build a tower, doth not first sit down and count the cost, whether he have wherewith to complete it?" *Luke 14:28, ASV.*

One of the most painful lessons I have had to learn had to do with good friends and work projects. I had a friend since high school that I discovered lived near us when we moved to our current home. We made arrangements to get together as families and did so on several occasions. He was a building contractor, and I needed to have dormers put on the second floor of our home. When I asked him about it, he said it was a job that he and I working together could knock out in a day. This was the first construction project I had ever hired out, so I was remembering what he had said about doing the job in two days as an estimate. I gave him the green light, and he and his crew of three guys showed up one morning. I was thinking they would be done in one day or at most two days. After the first day, the equipment was up and one hole was cut in the roof. After two days, one of the dormers was framed in and things were moving along. We had to go away for a few days, so I expected to return and find all the work completed. When we drove into the driveway, indeed it was almost complete, except for two guys very slowly putting the finishing touches on the siding. It was painful to watch.

The end of the story is that he presented me with a bill that was about five times what I had in my mind. If I had received at least an estimate beforehand, I would have known whether or not I could have afforded this work and would have had a reasonable figure in my mind. And if I hadn't witnessed the two turtles finishing up the job, I would have paid the tab more willingly. Needless to say, I wasn't happy about the bill, and we had an uncomfortable and unpleasant discussion. He lowered the price somewhat and I paid him that amount. Shortly thereafter, I sent him money to pay the exact amount of the original bill, but despite repeated phone calls, we have never spoken since. I still grieve over this, but it could have been avoided if I had followed basic principles of common sense that are revealed in the Bible.

PRAYER

Thank You for Your Word which is a "lamp unto my feet, and light unto my path" *Psalm 119:105.* Amen.

LESSON 28

Dirty Hands

There is merit in working with one's hands. It is a good thing to have a trade of some sort, whether as a painter, carpenter, electrician, mechanic, or something else. Not only will it help you through dry spells when you may be out of work at your regular job, but it will also serve as a useful skill to bless others.

"And that ye study to be quiet, and to do your own business, and to work with your own hands, as we commanded you; that ye may walk honestly toward them that are without, and that ye may have lack of nothing" *1 Thessalonians 4:11–12.*

"Let him that stole steal no more: but rather let him labour, working with his hands the thing which is good, that he may have to give to him that needeth" *Ephesians 4:28.*

Paul could write these words because he was a tentmaker and practiced what he preached. "And because he was of the same trade, he abode with them, and they wrought, for by their trade they were tentmakers" *Acts 18:3, ASV.*

As a tentmaker, he was able to supply his own needs. "Ye yourselves know that these hands ministered unto my necessities, and to them that were with me" *Acts 20:34, ASV.*

"For ye remember, brethren, our labor and travail: working night and day, that we might not burden any of you, we preached unto you the gospel of God. " *1 Thessalonians 2:9, ASV.*

When I was a senior in high school, my youth group participated in a short-term mission project to one of the poorest counties in the U.S. While working with needy families in rural Kentucky, I was able to use many of the carpentry skills I had learned while working on a construction crew at the beginning of the summer.

After my one summer as a carpenter's helper, I became a house painter. Not only did I learn a practical skill, but I was also able to earn money for college and seminary. There have been many work projects on church buildings and at church members' homes where my painting ability has come in handy. I have been in charge of painting everything from an all-purpose room to the exterior of a large frame building that used over 125 gallons of paint. Not only did both jobs meet specific needs, but they saved the local body of believers thousands of dollars as well.

Upon graduation from seminary, I moved to a small church to work as the assistant to the pastor. Before I was ordained as the pastor, God led me to work at a local cabinet shop. That job helped me in many ways. Not only did I acquire more skills with my hands, but working in that environment also helped me to relate to a few of the church members who had similar blue-collar jobs. The foreman at that shop is one of my best friends and became an elder at the church.

Certainly the worthy woman described in the last chapter of Proverbs exhibits these qualities of diligently working with her hands.

"She seeks wool, and flax, and works willingly with her hands. She is like the merchants' ships; she brings her food from afar. She rises also while it is yet night, and gives meat to her household, and a portion to her maidens" *Proverbs 31:13–15, AKJV.*

"She stretcheth out her hand to the poor; yea, she reacheth forth her hands to the needy" *Proverbs 31:20.*

"She looks well to the ways of her household, and does not eat the bread of idleness" *Proverbs 31:27, ESV.*

In this age of technology, I believe there is still value in getting your hands dirty in good old-fashioned labor. I still prefer to cut the grass and take care of the maintenance of our home myself. God assigned the task of tending the Garden of Eden to Adam, and perhaps that is why we still like to work with our hands.

PRAYER

Father, knowing that all scripture is inspired and profitable for teaching that we may be complete in every good deed, help us to know how to apply these scriptures in the twenty-first century in which we live. Amen.

LESSON 29

The Lord Bless and Keep You

"The blessing of the Lord, it makes rich, and he adds no sorrow with it" *Proverbs 10:22, AKJV.*

I know of nothing else I desire more, when setting out to work with God, than His blessing. You can have the best idea and be very diligent in applying all the biblical principles we have studied, but God's blessing is what you must have to succeed. Today we use the expression "God bless you" as a nice thought, or we say it when someone sneezes. But the blessing of God was nothing to be sneezed at by the patriarchs Isaac, Jacob, Esau, and Joseph. God's blessing was a very real and tangible thing. It made all the difference to them. As we look through the 20/20 hindsight of history, it was the blessing passed down from father to son that set Israel apart from the other nations. The blessing given to Abraham was passed on to Isaac, then to Jacob and his twelve sons.

When the disciples asked Jesus how to pray, one of the phrases in the Lord's Prayer referred to our daily bread: "Give us this day our daily bread" *Matthew 6:11.* This prayer should be a regular part of our lives, because regardless of our salary or our inheritance, ultimately it is God who provides for us. We need His blessing on our homes for our daily provisions.

When we were working at a church with very little visible means of support, it was God's blessing that provided for our needs. The church we were a part of provided room and board and a vehicle for church use, but the rest was between us and our heavenly Father. For the four years when we didn't have any regular income, we learned what it meant to have God provide for our needs. He was faithful. But whether we have a salary or not, we still need to pray for God's daily sustenance. All of the Christian life is lived by faith. And faith is the conviction of things not seen. The work of God is not often seen, but it is still real.

There is a curious expression in Haggai about a bag with holes. "You have planted much but harvest little. You eat but are not satisfied. You drink but are still thirsty. You put on clothes but cannot keep warm. Your wages disappear as though you were putting them in pockets filled with holes!" *Haggai 1:6, NLT.*

These people worked hard, but it was like putting their money into a bag with holes. Their needs were not being met. In looking back over my life, I can testify that God met my family's needs, but I don't know how. I just know that at the end of the year we were not in debt and God had provided our daily bread. That is living by faith. As we obey God and put our trust in Him, He knows how to plug the holes in our bag and bless our little pot of oil as Elisha did for the widow in 2 Kings 4. He takes good care of His children.

We need God's touch on our health, our ability to work, our vehicles, etc. But it is easier to keep our eyes on God when there is no regular source of income. God knew this when He spoke to the children of Israel in Deuteronomy.

"And thou shalt remember all the way which the Lord thy God led thee these forty years in the wilderness, to humble thee, and to prove thee, to know what was in thine heart, whether thou wouldest keep his commandments, or no. And he humbled thee, and suffered thee to hunger, and fed thee with manna, which thou knewest not, neither did thy fathers know; that he might make thee know that man doth not live by bread only, but by every word that proceedeth out of the mouth of the Lord doth man live. Thy raiment waxed not old upon thee, neither did thy foot swell, these forty years... When thou hast eaten and art full, then thou shalt bless the Lord thy God for the good land which he hath given thee... Lest when thou hast eaten and art full, and hast built goodly houses, and dwelt therein; and when thy herds and thy flocks multiply, and thy silver and thy gold is multiplied, and all that thou hast is multiplied; then thine heart be lifted up, and thou forget the Lord thy God, which brought thee forth out of the land of Egypt" *Deuteronomy 8:2–4, 10, 12–14.*

So how do you get and maintain God's blessing? Much of what we have already said outlining God's principles applies here. Being honorable, honest, and diligent, praying over decisions and practices, returning a tenth; keeping a day of rest; and honoring your wife as the weaker vessel are all ways of working with God. When you conduct your affairs in a way that honors God and is in accordance with the precepts spelled out in His Word, you can expect God to work with you. A missionary once said, "God's work done in God's way will never lack God's supply." All work is God's work when done unto Him. Also, don't ever forget to just plain ask for God to bless the work of your hands. I had a faithful mentor pray for our business every day for several years, and other friends and pastors have done similarly. I value their prayers. I value God who answers them.

PRAYER

God, bless us in working with You to provide for our families. Give us this day our daily bread. Cause us by Your Spirit to work in such a way that You continue to bless us. Amen.

A DAY OF REST

One tangible expression of God's blessing is a day of rest.

"And on the seventh day God finished his work which he had made; and he rested on the seventh day from all his work which he had made. And God blessed the seventh day, and hallowed it; because that in it he rested from all his work which God had created and made" *Genesis 2:2–3, ASV.*

"Remember the sabbath day, to keep it holy. Six days shalt thou labor, and do all thy work; but the seventh day is a sabbath unto Jehovah thy God: in it thou shalt not do any work, thou, nor thy son, nor thy daughter, thy man-servant, nor thy maid-servant, nor thy cattle, nor thy stranger that is within thy gates: for in six days Jehovah made heaven and earth, the sea, and all that in them is, and rested the seventh day: wherefore Jehovah blessed the sabbath day, and hallowed it" *Exodus 20:8-11.*

There are two thoughts revealed in these scriptures that are extremely obvious and often overlooked: work six days and rest one day. You do not need to be able to read the Bible in the original languages to understand this. Notice also that even though Moses was the first to codify this formula, it was God who first popularized this approach.

Working six days is an exhortation to diligence, which we discussed in a previous section. Resting for one day, as God did, is a prescription for sanity and a wonderful antidote to the culture of stress in which we live. God set the example. He worked for six days and then rested for one day. Can we improve upon this model? I think not.

Here are a few thoughts for you to ponder. First determine in your own home which day you will observe.

"One man esteemeth one day above another: another esteemeth every day alike. Let each man be fully assured in his own mind. He that regardeth the day, regardeth it unto the Lord" *Romans 14:5–6, ASV.*

I get two insights from these verses. First, be fully persuaded in your own mind. Search the scriptures. There are over 200 references to the Sabbath in scripture, and studying them is enlightening. Seek God for wisdom on this issue. Then when you have arrived at a decision, apply the second insight, which is to give your brothers and sisters in the faith some space. I am fully persuaded that Saturday is the seventh day, the Sabbath. I am also convinced Sunday is the first day of the week, or the Lord's Day. But if others don't agree, I am not going to make this a bone of contention; rather I am going to keep these days unto the Lord.

When you have arrived at a decision before God, then practice it and help ohters to really rest one day each week. This is how God made us, to work six days and rest one day per week. Emergencies and unique circumstances will arise, but if you make this a habit in your home and business, you will be working with the Creator. How you keep it is between you and God, but when properly practiced under Christ and in the rest that only He can provide, it will be a source of weekly refreshment and a regular window of opportunity to spend time with God and your family.

"There remaineth therefore a sabbath rest for the people of God" *Hebrews 4:9, ASV.*

"But now we have been discharged from the law, having died to that wherein we were held; so that we serve in newness of the spirit, and not in oldness of the letter" *Romans 7:6, ASV.*

"For freedom did Christ set us free: stand fast therefore, and be not entangled again in a yoke of bondage" *Galatians 5:1, ASV.*

PRAYER

Help us to set aside one day each week for our necessary rest. Inspire us to do so unto the Lord in newness of the spirit and not in oldness of the letter. Amen.

LESSON 30

Thanks Giving

"Every good gift and every perfect gift is from above, coming down from the Father of lights, with whom can be no variation, neither shadow that is cast by turning" *James 1:17, ASV.*

"In everything give thanks: for this is the will of God in Christ Jesus to you-ward" *1 Thessalonians 5:18, ASV.*

The focus of this book is how to be a wise steward. We have examined how to be a careful spender, a diligent provider, and a wise planner. Our goal has been to follow Jesus in every area of our lives, including our spending. Let us also be thankful.

For all of us who have received so much from our good God, the best and most appropriate thing we can do is recognize that "every good and perfect gift comes from above" and say, "Thank you." It is natural, when we have needs, to work and pray to see those needs met. But when we are fed, clothed, and cared for in a variety of ways, let's not forget who helped us.

I can't think of a more awful malady than leprosy. It is a terrible disease that causes physical suffering and social ostracism. Luke records the story of how ten lepers approached

Jesus and lifted up their voices for help. Their appeal was heard and each was miraculously healed of this terrible disease. One man returned to give glory to God and give Him thanks. Only one. This seems to be pretty normal behavior for humanity in general. We receive so much and yet forget to give God thanks.

"And as he entered into a certain village, there met him ten men that were lepers, which stood afar off: and they lifted up their voices, and said, Jesus, Master, have mercy on us. And when he saw them, he said unto them, Go shew yourselves unto the priests. And it came to pass, that, as they went, they were cleansed. And one of them, when he saw that he was healed, turned back, and with a loud voice glorified God, and fell down on his face at his feet, giving him thanks: and he was a Samaritan. And Jesus answering said, Were there not ten cleansed? but where are the nine? There are not found that returned to give glory to God, save this stranger" *Luke 17:12–18.*

As a husband and father, I am committed to caring for my wife and children. I like working for them. I want to provide for their needs. They are my family, and I love them. But I really appreciate it when they thank me for what I do for them. Whether it is taking them to play mini-golf, buying them a meal at a restaurant, or taking them on a vacation, it always blesses me when they say thanks. On more than one occasion, my wife has come into my office when I was working and thanked me for working for my family. She doesn't need to do this, as I will probably do it anyway, but it sure does make my work more enjoyable.

"And let the peace of Christ rule in your hearts, to the which also ye were called in one body; and be ye thankful" *Colossians 3:15, ASV.*

PRAYER

Dear Father, thank You. Amen.

A FINAL NOTE

This book was 31 years in the making and living and two years in the writing. I hope it encourages you and provides some food for thought and reflection. I also pray it will not be a source of discouragement or condemnation.

"There is therefore now no condemnation to them which are in Christ Jesus, who walk not after the flesh, but after the Spirit" *Romans 8:1.*

If you are just getting started in this area of letting Christ be the Lord of your wallet, don't try to do everything at once. Let God lead you in how to incorporate these principles into your life. Walk with Him and walk in the light that He gives. His yoke is easy and His burden is light.

"Come unto me, all ye that labour and are heavy laden, and I will give you rest. Take my yoke upon you, and learn of me; for I am meek and lowly in heart: and ye shall find rest unto your souls. For my yoke is easy, and my burden is light" *Matthew 11:28–30.*

Remember the grace of God, the mercy of God, and the forgiveness of God. He is the source of new beginnings and fresh hope. He is the source of encouragement. Regardless of your past, He is the one who makes all things new!

And he that sat upon the throne said, Behold, I make all things new" *Revelation 21:5.*

Our enemy, the devil, is the author of discouragement and despair.

"And I heard a loud voice saying in heaven, Now is come salvation, and strength, and the kingdom of our God, and the power of his Christ: for the accuser of our brethren is cast down, which accused them before our God day and night" *Revelation 12:10.*

Dark thoughts of "What's the use?" and hopelessness come from the pit. New life, fresh faith, high hopes, and good courage come from Jesus, the source of light and love.

"The thief cometh not, but for to steal, and to kill, and to destroy: I am come that they might have life, and that they might have it more abundantly" *John 10:10.*

May God bless each of you in following Him,

Steve Demme

APPENDIX A

Hezekiah's Open Checkbook

Keep your checkbook close to your chest. Only God, you, and your spouse need to know how you are doing financially and how much money you have (or don't have).

"He that guardeth his mouth keepeth his life; but he that openeth wide his lips shall have destruction" *Proverbs 13:3, ASV.*

There is a penetrating example of this principle in the life of one of the finest kings in the history of the southern kingdom of Judah. After his sickness, Hezekiah received a visit from some Babylonian princes.

"At that time Berodach-baladan the son of Baladan, king of Babylon, sent letters and a present unto Hezekiah; for he had heard that Hezekiah had been sick. And Hezekiah hearkened unto them, and showed them all the house of his precious things, the silver, and the gold, and the spices, and the precious oil, and the house of his armor, and all that was found in his treasures: there was nothing in his house, nor in all his dominion, that Hezekiah showed them not" *2 Kings 20:12–13, ASV.*

Hezekiah shared all and left nothing hidden. I don't know all the implications of this account, but it continues with a visit from God's man on the spot, the prophet Isaiah.

"Then came Isaiah the prophet unto king Hezekiah, and said unto him, What said these men? and from whence came they unto thee? And Hezekiah said, They are come from a far country, even from Babylon. And he said, What have they seen in thy house? And Hezekiah answered, All the things that are in my house have they seen: there is nothing among my treasures that I have not showed them. And Isaiah said unto Hezekiah, Hear the word of Jehovah. Behold, the days come, that all that is in thy house, and that which thy fathers have laid up in store unto this day, shall be carried to Babylon: nothing shall be left, saith Jehovah" *2 Kings 20:14–17, ASV.*

As I said before, I don't know all that was going on here between Hezekiah and God, but there seems to be a direct correlation between his revealing all and the Babylonians taking all. Did he incite the princes of Babylon to covetousness by displaying all of his treasures? I don't know. But I do know that these historical accounts are written for you and me. They are for our instruction.

"Now all these things happened unto them for examples: and they are written for our admonition, upon whom the ends of the world are come" *1 Corinthians 10:11.*

It doesn't do any good to show your material blessings. It may cause your brother to covet, which may have been the case for the Babylonians. Perhaps the reason, deep down in your heart of hearts, is not to bring glory to God for His blessings, but to show how much stuff you have, which is pride. Pride always goes before a fall and is the antithesis of humility, which is the appropriate and God-honoring characteristic of a believer.

"Pride goeth before destruction, and a haughty spirit before a fall. Better it is to be of a lowly spirit with the poor, than to divide the spoil with the proud" *Proverbs 16:18–19, ASV.*

PRAYER

We trust You, God, to help us have discretion and wisdom in sharing our God- given blessings with others. Amen.

APPENDIX B

Bribes for the Blind

"A wicked man takes a gift out of the bosom to pervert the ways of judgment" *Proverbs 17:23, AKJV.*

"And thou shalt take no bribe: for a bribe blindeth them that have sight, and perverteth the words of the righteous." *Exodus 23:8, ASV.*

The obvious application of these scriptures is to not take a bribe, because it will cloud your sense of right and wrong. I have been offered a bribe only once that I can remember, and that was to change a student's grade. My brother was interested in buying a piece of property, and I knew that the father of one of my students sold real estate. So one day we met the man, and he offered to show us a few parcels of land. On the way, he asked me to drive in his pickup truck while my brother and his wife followed in their car. On the way, he offered me an acre of land if I would change his son's grade from an F to a D. I thought the man was kidding, so I laughed out loud, only to realize I was the only one laughing. I didn't accept the offer, and the student flunked the class. Now I don't know how many situations you will find yourself in where you will have an opportunity to put these verses into practice, but there is another area that is similar, and that is receiving free gifts.

I always heard that nothing is free. In the case of merchandising, I am tempted to agree. When you receive a perk, like a set of kitchen knives, for attending a sales presentation, it engenders a sense of obligation on your part. If you test drive a new car and they give you a free DVD or a voucher for a free meal, the gift does make you more inclined to purchase the product. If it didn't have this effect, the gift wouldn't be used. Our human nature is wired so that when we are given something for nothing, we feel that we owe the giver something in return. Perhaps it even blinds us a little, as Exodus 23:8 states above. I find it easier to not accept any gift in the first place. Perhaps you are stronger than I am in this matter, but this seems to be the most effective way. Of course, if you go to see a timeshare presentation and get a gift, the time you invested in listening to the spiel is worth something, like a set of knives!

A final word to the wise: be careful of accepting free things that may blind your judgment or create a false sense of obligation.

PRAYER

Save us from taking bribes or gifts that will blind our judgment and hinder our ability to make a clear, thoughtful decision. Amen.

Scripture Index